Florian Pausch

Spatial audio reproduction for hearing aid research

System design, evaluation and application

Logos Verlag Berlin GmbH

Aachener Beiträge zur Akustik

Editors:
Prof. Dr.-Ing. Janina Fels
Prof. Dr. rer. nat. Michael Vorländer
Institute for Hearing Technology and Acoustics
RWTH Aachen University
52056 Aachen
www.akustik.rwth-aachen.de

Bibliographic information published by the Deutsche Nationalbibliothek

The Deutsche Nationalbibliothek lists this publication in the Deutsche Nationalbibliografie;
detailed bibliographic data are available in the Internet at http://dnb.d-nb.de.

D 82 (Diss. RWTH Aachen University, 2021)

ISBN 978-3-8325-5461-3
ISSN 2512-6008
Vol. 36

Logos Verlag Berlin GmbH
Georg-Knorr-Str. 4, Geb. 10,
D-12681 Berlin
Tel.:   +49 (0)30 / 42 85 10 90
Fax: +49 (0)30 / 42 85 10 92
http://www.logos-verlag.de

To my parents

Abstract

Hearing loss (HL) has multifaceted negative consequences for individuals of all age groups. Despite individual fitting based on clinical assessment, consequent usage of hearing aids (HAs) as a remedy is often discouraged due to unsatisfactory HA performance. Consequently, the methodological complexity in the development of HA algorithms has been increased by employing virtual acoustic environments which enable the simulation of indoor scenarios with plausible room acoustics. Inspired by the research question of how to make such environments accessible to HA users while maintaining complete signal control, a novel concept addressing combined perception via HAs and residual hearing is proposed. The specific system implementations employ a master HA and research HAs for aided signal provision, and loudspeaker-based spatial audio methods for external sound field reproduction. Systematic objective evaluations led to recommendations of configurations for reliable system operation, accounting for perceptual aspects. The results from perceptual evaluations involving adults with normal hearing revealed that the characteristics of the used research HAs primarily affect sound localisation performance, while allowing comparable egocentric auditory distance estimates as observed when using loudspeaker-based reproduction. To demonstrate the applicability of the system, school-age children with HL fitted with research HAs were tested for speech-in-noise perception in a virtual classroom and achieved comparable speech reception thresholds as a comparison group using commercial HAs, which supports the validity of the HA simulation. The inability to perform spatial unmasking of speech compared to their peers with normal hearing implies that reverberation times of 0.4 s already have extensive disruptive effects on spatial processing in children with HL. Collectively, the results from evaluation and application indicate that the proposed systems satisfy core criteria towards their use in HA research.

Kurzfassung

Hörverlust (HV) hat vielfältige negative Auswirkungen auf Menschen aller Altersgruppen. Trotz individueller Anpassung auf Basis von klinischen Verfahren werden Hörgeräte (HG) als Gegenmaßnahme wegen unzureichender Leistung von Betroffenen oft nicht akzeptiert. Infolgedessen wurde die methodische Komplexität in der Entwicklung von HG-Algorithmen durch die Verwendung von virtuellen akustischen Umgebungen, welche die Simulation von Innenraumszenarien mit plausibler Raumakustik ermöglichen, erhöht. Um HG-Träger*innen solche Umgebungen zugänglich zu machen, wird ein neuartiges Konzept zur Berücksichtigung der kombinierten Wahrnehmung über HG und das Resthörvermögen vorgeschlagen. Die Systemimplementierungen verwenden eine Softwareplattform für HG-Algorithmen in Kombination mit Forschungs-HG für die Wiedergabe der HG-Signale, und lautsprecherbasierte räumliche Wiedergabemethoden für die externe Schallfeldwiedergabe. Für einen zuverlässigen Systembetrieb werden unter Berücksichtigung perzeptiver Anforderungen Empfehlungen zu Systemkonfigurationen abgegeben. Untersuchungen mit normalhörenden Erwachsenen zeigen, dass die Eigenschaften der Forschungs-HG primär die Schalllokalisation beeinflussen, jedoch vergleichbare egozentrische auditive Entfernungsschätzungen wie bei lautsprecherbasierter Wiedergabe ermöglichen. Zur Demonstration der Systemanwendbarkeit wurde die Sprachwahrnehmung von Schulkindern mit HV bei Verwendung der Forschungs-HG in einem virtuellen Klassenzimmer in Präsenz von Störsprecherinnen getestet. Ähnliche Sprachverständlichkeitsschwellen wie jene einer Vergleichsgruppe mit kommerziellen HG bekräftigen die Plausibilität der HG-Simulation. Die Unfähigkeit der räumlichen Demaskierung von Sprache deutet darauf hin, dass bereits Nachhallzeiten ab 0,4 s die räumliche Verarbeitung von Kindern mit HV erheblich beeinträchtigen. Insgesamt zeigen die Evaluierungs- und Anwendungsergebnisse, dass die vorgeschlagenen Systeme wesentliche Kriterien für ihren Einsatz in der HG-Forschung erfüllen.

Contents

Glossary

Notation

$S(k)$	Frequency spectrum of the discrete-time signal $s(n)$
$\mathbf{M}$	Matrix
$\mathbf{v}$	Vector in matrix notation
$s(n)$	Discrete-time signal
$s(t)$	Continuous-time signal

Operators

$(\cdot)!$	Factorial
$(\cdot)^{\mathsf{T}}$	Transpose of a vector or matrix
$\Re\{\cdot\}, \Im\{\cdot\}$	Real and imaginary part of a complex number
$*$	Linear convolution
$\circ$	Element-wise multiplication
$\|\cdot\|_2$	L2 norm of a vector
$\|\cdot\|$	Absolute value; magnitude of a complex number
$\mathcal{DFT}_{\mathrm{K}}\{\cdot\}$	K-point discrete Fourier transform
$\mathcal{DFT}_{\mathrm{K}}^{-1}\{\cdot\}$	Inverse K-point discrete Fourier transform
$\log(\cdot)$	Common logarithm (base 10)
$\arg\{\cdot\}$	Argument of a complex number
$\mathrm{diag}\{\cdot\}$	Square diagonal matrix
$\overline{(\cdot)}, (\cdot)^{\mathsf{H}}$	Conjugate-complex transpose (Hermitian transpose)
$(\cdot)^{-1}$	Matrix inverse

Signals and functions

$P_{n,\|m\|}$	Associated Legendre function of the first kind of order n and degree m
$E_{\mathrm{S}}(t)$	Energy decay curve obtained from Schroeder's backward integration method
$P(k)$	Sound pressure signal in frequency domain (sound pressure spectrum)
$Y_n^m(\theta, \varphi)$, $\boldsymbol{y}_{\mathcal{N}}$, $\mathbf{Y}_{\mathcal{N}}$	Spherical harmonic basis function, vector and matrix (up to a truncation order $\mathcal{N}$)
$\delta(n)$	Unit impulse
$b_{\{\mathrm{L,R}\}}(n)$, $B_{\{\mathrm{L,R}\}}(k)$, $\mathbf{b}$	Left (L) and right (R) binaural input signals with corresponding discrete Fourier transform spectra
$e_{\{\mathrm{L,R}\}}(n)$, $E_{\{\mathrm{L,R}\}}(k)$, $\mathbf{e}$	Left (L) and right (R) binaural signals at ear level and corresponding discrete Fourier transform spectra
$h(n)$, $H(k)$	Impulse response of a linear time-invariant system and corresponding discrete Fourier transform spectrum (transfer function)
$s(n)$, $S(k)$	Exponentially swept sine in time and frequency domain
$x(n)$, $X(k)$	Input signal of a linear time-invariant system and corresponding discrete Fourier transform spectrum
$y(n)$, $Y(k)$	Output signal of a linear time-invariant system with corresponding discrete Fourier transform spectrum

Symbols

A_{T}	Difference between the equivalent sound absorption areas of the reverberation chamber with and without test object	m^2
C'_{50}	Intelligibility-weighted composite clarity index for speech	dB
C_{50}	Clarity index for speech	dB
C'_{80}	Averaged composite clarity index for music	dB
C_{80}	Clarity index for music	dB

D_0	Far-field directivity of a sound source referenced to its on-axis directivity	
D_{50}	Speech definition ("Deutlichkeit")	
F	Fisher's ratio	
$L_{W,Z}$	Sound power level with flat (Z) frequency weighting	dB
$L_{Z,S,max}$	Maximum sound pressure level with flat (Z) frequency weighting and slow (S) time weighting	dB
$L_{Z,eq}$	Equivalent continuous sound pressure level with flat (Z) frequency weighting	dB
R^2	Coefficient of determination, representing the proportion of the explained variance.	
R	Loudspeaker array radius	m
$T_{30,mid}$	Mean mid-frequency reverberation time, representing T_{30} averaged across 500 Hz and 1 kHz	s
T_{30}	Reverberation time, estimated by evaluating the normalised energy decay curve	s
T_s	Sampling period	s
V	Volume	m^3, ℓ
Ξ	Auditory horizon	m
α_S	Ratio of the equivalent sound absorption area of a test object to the area of the test object	
α_p	Practical sound absorption coefficient	
β	Tikhonov regularisation factor	
$\boldsymbol{\theta}$	Direction vector	
$\boldsymbol{r}_E$	Energy vector	
$\boldsymbol{r}_V$	Velocity vector	
$\check{\mathbf{H}}$	Matrix of playback transfer functions	
χ^2	Test statistic of the likelihood-ratio test	
ϵ	Error with indices referring to φ, ϑ/θ and γ (angular errors), or r/R (radial error)	deg, m
η	Loss factor of a material	
γ	Great circle central angle	deg
$\hat{d}$	Estimated egocentric distance	m
κ	Wave number, $\kappa = 2\pi f/c$	m^{-1}
λ	Wavelength, $\lambda = c/f$	m
$\mathbf{C}$	Matrix of acoustic crosstalk cancellation filters	
$\mathbf{I}$	Identity matrix	
Q	Number of distances	

$\mathcal{N}$	Truncation order of the spherical harmonic expansion	
m	Spherical harmonic degree	
n	Spherical harmonic order	
μ	Arithmetic mean of a sample	
σ^2	Residual variance (within-condition variance) in linear mixed-effect models	
σ_{se}	Standard error of μ	
σ	Standard deviation of a sample with Bessel's correction	
τ_{00}	Mean random effect variance (between-condition variance) in linear mixed-effect models	
K	Number of frequency bins	
M	Signal length in time domain	samples
N'	Number of involved loudspeakers	
N	Total number of loudspeakers	
θ	Zenith angle, $0° \leq \theta \leq 180°$	deg
η_{p}^2	Measure of effect size for analysis-of-variance models	
ε	Greenhouse-Geisser estimate of sphericity	
φ	Azimuth angle, $0° \leq \varphi < 360°$	deg
ϑ	Elevation angle, $-90° \leq \varphi \leq 90°$	deg
ξ	Linear compression parameter in the compressive power function egocentric distance model	
ζ	Non-linear compression parameter in the compressive power function egocentric distance model	
$a_n, \boldsymbol{a}_{\mathcal{N}}$	Spherical harmonic-order dependent weight(s), scalar and vector	
c	Speed of sound	m/s
d	True egocentric distance	m
f_0	Resonance frequency of a system	Hz
f_{S}	Schroeder frequency	Hz
f_{s}	Sampling rate	Hz
f	Frequency	Hz
$g, \mathbf{g}$	Amplitude loudspeaker weights for panning-based reproduction	
i, l, q	Indices	
j	Imaginary unit, $j = \sqrt{-1}$	
k	K-point discrete Fourier transform index	
n	Discrete-time index	

p	Probability value (p-value)	
r	Radial distance	m
t_{mp95}	Perceptual mixing time	s
t	Time; test statistic of the t-test; order of a spherical t-design layout	s; (); ()
w	Voronoi weights	

Acronyms

4-PTA	four-frequency pure tone average
A/D	analog-to-digital
AIC	Akaike information criterion
AllRAD+	improved all-round ambisonic decoder
ANL	ambient noise level
ANOVA	analysis of variance
ASIO	Audio Stream Input/Output
BRIR	binaural room impulse response
BTE	behind-the-ear
CI	confidence interval
CS	channel separation
CTC	crosstalk cancellation
D/A	digital-to-analog
DFT	discrete Fourier transform
DRR	direct-to-reverberant ratio
DS	direct sound
DSP	digital signal processor
EDC	energy decay curve
EEL	end-to-end latency
EQ	equalisation
FIR	finite impulse response
GUI	graphical user interface
HA	hearing aid
HAA	hearing aid auralisation
HARIR	hearing aid-related impulse response
HARRIR	hearing aid-related room impulse response
HARTF	hearing aid-related transfer function
HL	hearing loss
HOA	higher-order Ambisonics
HRIR	head-related impulse response

HRTF	head-related transfer function
IACC	interaural cross correlation
ICC	intra-class correlation
ILD	interaural level difference
IR	impulse response
ITD	interaural time difference
JND	just noticeable difference
LiSN-S	Listening in Spatialized Noise–Sentences
LME	linear mixed-effect
LTI	linear time-invariant
MDAP	multiple-direction amplitude panning
MHA	master hearing aid
MobiLab	mobile laboratory
MPANL	maximum permissible ambient noise level
n.s.	not significant
NH	normal hearing
PTA	pure tone audiometry
RITE	receiver-in-the-ear canal
RMS	root mean square
RSS	real sound source
RT	reverberation time
SAD	sampling ambisonic decoder
SD	spectral difference
SE	standard error
SH	spherical harmonic
SiN	speech-in-noise
SNR	signal-to-noise ratio
SPL	sound pressure level
SRI	sound reduction index
SRM	spatial release from masking
SRT	speech reception threshold
SWL	sound power level
TOA	time of arrival
VAE	virtual acoustic environment
VBAP	vector base amplitude panning
VBIP	vector base intensity panning
VDL	variable delay line
VR	virtual reality
VSS	virtual sound source

1

Introduction

Listening, understanding and communicating are essential skills for social interaction in everyday life. Hearing loss (HL) may have severe consequences on these skills and affect individuals by hindering language development and learning success already at an early age (Stevenson et al., 2010; Richardson, Long, & Woodley, 2004). According to the World Health Organization (2021), about 5 % of the world population, amounting to 432 million adults and 34 million children, suffer from disabling HL – a number that is estimated to grow to 700 million people by 2050. Depending on the individual predisposition, affected children are likely to withdraw from socialisation processes and sometimes suffer from compromised mental and general health (Levy-Shiff & Hoffman, 1985; Hogan et al., 2011; Stevenson et al., 2011; Patel et al., 2021). Similar negative consequences may also apply to adults and elderly people, especially those who refuse to acknowledge their HL and appropriate treatment, leading to an increased risk of isolation (Mick, Kawachi, & Lin, 2014; Shukla et al., 2020), depression (Boi et al., 2012; Mener et al., 2013), as well as memory problems and dementia (Wingfield, Tun, & McCoy, 2005; Moorman, Greenfield, & Lee, 2021).

Selectively listening to a target talker while ignoring masking and distracting noise sources – commonly referred to as the "cocktail party effect" (Cherry, 1953) – is among the most important abilities for successful communication. The outcomes of behavioral studies in the field of speech-in-noise (SiN) perception corroborate inferior unmasking performance of target speech in individuals with HL across age groups, compared to age-matched individuals with normal hearing (NH) (Cameron, Glyde, & Dillon, 2011; Bronkhorst, 2000). From a historical

---

Related publication:
Pausch et al., 2018b.

perspective, cupping the hand behind the ear can be considered maybe the first natural and still extensively used technique to increase speech intelligibility in such situations. Besides acting as a visual sign for the conversational partner to speak up, the cupped hand creates acoustic resonances between 1 and 3 kHz that are beneficial for speech understanding, and enhances directivity to increase the signal-to-noise ratio (SNR) between the target talker and detrimental noise sources (Uchanski & Sarli, 2019). Possibly inspired by the shape of the ears in animals, people also experimented with mechanical hearing aids (HAs) in the form of hearing trumpets, allowing to further increase directivity and tailor the resonance frequencies. Such devices had a multifaceted appearance, including chairs with two attached horns, representing probably one of the first approaches to provide binaural amplification (Stephens & Goodwin, 1984; Hawley, Litovsky, & Culling, 2004). Electric HAs were implemented by the end of the 19th century, initiated by the invention of the telephone, and since then continuously miniaturised by incorporating advancing technologies such as integrated circuit designs (Mills, 2011). Unobtrusive and miniaturised designs contributed significantly to a higher acceptance rate of HAs in view of the stigma of HL (David & Werner, 2016). While the parameters of the first analogue devices were adjusted using potentiometers, rapid progress in digital technology soon enabled to program analogue devices digitally, and to develop completely digitised devices with on-board digital signal processors (DSPs) (Kates, 2008). Nowadays, assistive listening devices, such as HAs, represent an effective technical intervention tool capable of partially restoring impaired hearing and helping to overcome the aforementioned consequences of HL (Chisolm et al., 2007; Tomblin et al., 2014). Various sophisticated HA algorithms have been specifically designed to address the deficiencies in affected individuals (Hamacher et al., 2005; Kates, 2008). More recently, access to the devices has become even more convenient by the availability of over-the-counter HAs (Warren & Grassley, 2017) and self-fitting approaches (Keidser & Convery, 2016). Premium devices sometimes feature additional acoustic scene classification algorithms (Yellamsetty et al., 2021), utilising machine learning approaches such as deep learning (Vivek, Vidhya, & Madhanmohan, 2020).

Despite the major advances in technology, extensive surveys reported that HA users still complain about poor performance of their devices, particularly in noisy situations like demanding indoor communication settings with unfavourable room acoustics (Hougaard & Ruf, 2011) – although there seems to be a trend towards higher satisfaction rates (Bisgaard & Ruf, 2017). Dissatisfaction may consequently lead to irregular device usage (Bertoli et al., 2009) and, if at all, only marginal improvement of the original problems. To some extent, the experienced performance deficiencies can be attributed to clinical assessment and fitting routines for the parametrisation of HAs, which may result in settings that are less effective in complex real-world situations (Cord et al., 2007; Timmer, Hickson, & Launer, 2018). Evaluating HA algorithms using oversimplified test scenarios also favours a gap between laboratory performance and reality. Significant advances in the field of acoustic virtual reality (Vorländer, 2020) have triggered research groups both in academia and industrial companies to utilise virtual acoustic environments (VAEs) for the design and improvement of HAs under more plausible conditions. The fact that reproduction of VAEs via headphones is not feasible and likely entails uncontrolled HA algorithm behaviour, let alone feedback issues, motivates the use of loudspeaker-based spatial audio reproduction (Minnaar, Favrot, & Buchholz, 2010; Grimm, Ewert, & Hohmann, 2015; Grimm, Kollmeier, & Hohmann, 2016; Oreinos & Buchholz, 2016). Investigations included, for example, assessing the real-world benefit of beamforming algorithms (Compton-Conley et al., 2004; Gnewikow et al., 2009), or aimed at perceptually validating SiN test results obtained in loudspeaker-based VAEs (Cubick & Dau, 2016). The availability of increased computational resources allowed to implement interactive low-latency listening scenarios using advanced room acoustic simulations in combination with highly efficient convolution algorithms (Noisternig et al., 2008; Pelzer et al., 2014; Mehra et al., 2015; Wefers, 2015; Schissler, Stirling, & Mehra, 2017).

This thesis contributes to the continued bridging of the aforementioned gap by exploring VAEs adapted for HA applications, rendered and reproduced by state-of-the-art simulation methods and spatial audio reproduction technologies, respectively. Novel binaural and hybrid concepts address the main research question of how to properly integrate HAs in the virtual scene, while retaining full control over HA algorithm parametrisation and the signals involved. The proposed systems are comparatively evaluated based on measurements and perceptual experiments, to be subsequently applied in a clinical experiment. Moreover, the flexibility and effectiveness of VAEs shall be demonstrated to further promote their application in HA research.

Thesis outline

A brief summary of the thesis structure is provided below per chapter. Chapter 2 introduces the required basics and fundamental concepts. At the beginning of Chapter 3, general application areas of VAEs are presented. Adaptations and extensions necessary to enable application in HA research are elaborated based on the proposed concept. After a compact introduction to current spatial audio reproduction methods, two specific implementation variants are discussed. Both approaches allow for transparent simulation and reproduction of signals as combined via research HAs and loudspeaker-based methods, either utilising binaural technology with a real-time processing backbone, or panning-based approaches as part of a static hybrid auralisation system. In Chapter 4, the two system variants become subject to objective evaluations based on different types of spatial rendering functions. Drawing on the results from a benchmark analysis of an example virtual indoor scene using the combined binaural real-time auralisation system, recommendations on simulation configurations for operation on conventional desktop computers are derived. Measurements on combined end-to-end latency (EEL) further provide insights on the reactiveness of the implemented system. Subsequently, results on the ideal and achievable channel separation (CS) in loudspeaker-based binaural reproduction under consideration of detrimental reflections is investigated. The errors in recreated spatial transfer functions obtained from different spatial audio reproduction methods are analysed at ear level in the presence of detrimental reflections. Chapter 5 investigates selected spatial audio quality parameters in the scope of two perceptual experiments involving adults with NH. In Experiment 1, differences in the localisation performance of virtual sound sources (VSSs), reproduced via headphones and the individual reproduction paths of the binaural real-time auralisation system, are compared to the localisation of real sound sources (RSSs). Experiment 2 assesses egocentric auditory distance perception in simulated room acoustics depending on variations in the specific implementation of the static hybrid auralisation system. Chapter 6 demonstrates the application of the binaural real-time auralisation system in a clinical experiment. Children with HL are tested for SiN perception in the presence of distracting talkers, while immersed in simulated room acoustics. The results are compared with those from an age-matched control group of children with NH to identify performance differences and discuss their reasons, addressing aspects of adverse room acoustics. Finally, the key findings are summarised and an outlook with ideas for future work is provided.

2

Fundamentals

This chapter provides concise descriptions of the fundamental concepts used in this thesis. The first part focuses on coordinate system conventions, digital signal processing basics, the measurement of impulse responses (IRs) of linear time-invariant (LTI) systems, such as rooms, room acoustic parameters, and source directivity. The second part deals with the human auditory system and threshold measurements, selected psychoacoustic topics, as well as HL and its perceptual effects. HAs and HA algorithm rationales are introduced and discussed in the scope of their rehabilitative purposes. Subsequently, binaural listening concepts are presented on the basis of directional transfer functions in listeners with and without HAs, introducing the most important localisation cues and differences between transfer function characteristics. The chapter ends with a short introduction to SiN tests and related terms to quantify SiN perception performance.

2.1 Coordinate systems

In this thesis, two different coordinate systems are used: a right-handed fixed *global* coordinate system and a right-handed variable *local* coordinate system, each with three orthogonal axes x, y and z, and x', y' and z', respectively, see Figure 2.1a. The positions of rigid bodies are represented by their current translational offsets to the origin of the global coordinate system in three-dimensional space, and accordingly defined by a triplet of Cartesian coordinates $[x \quad y \quad z]^\mathsf{T}$. Independent from the position, each rigid body has a current orientation. Po-

---

Related publications:

Pausch et al., 2018b; Pausch and Fels, 2020; Pausch, Doma, and Fels, 2021.

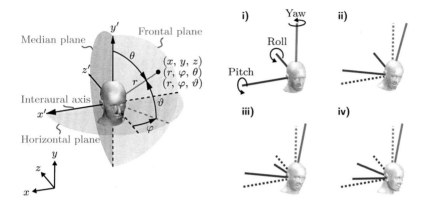

(a) Definition of the position of an example point in three-dimensional space by global Cartesian coordinates and local spherical coordinates. The listener-specific local coordinate system shows the interaural axis and the three planes, indicated but not limited by semicircular sectors. (Figure taken from Pausch and Fels, 2020, and adapted.)

(b) Example of an intrinsic rotation sequence with x', y' and z' axes color-coded in red, green and blue, respectively. Dashed and solid axes represent the original and sequentially rotated axes of the local coordinate system. **i)** Default orientation at (roll, pitch, yaw) = $(0°, 0°, 0°)$. **ii)** Intrinsic rotation by a roll angle of -15 deg. **iii)** Intrinsic rotation by a pitch angle of 10 deg. **iv)** Intrinsic rotation by a yaw angle of 20 deg.

Figure 2.1: Definitions of the global and local coordinate systems as per OpenGL conventions and corresponding object orientations. (Meshes created using Peyre, 2021.)

sition and orientation are collectively referred to by the term *pose* (Shapiro & Stockman, 2001).

In the specific local coordinate system associated with the indicated listener, three important planes (Blauert, 1997) are spanned by different local axis combinations: the horizontal plane (x' and z' axes, separating the upper and lower hemispheres), the median plane (y' and z' axes, separating the left and right half spaces), and the frontal plane (x' and y' axes, separating the front and rear half spaces). Although not generally the case, it is assumed for the purpose of simplicity and the time being that the axes of the local coordinate system intersect with the axes of the global coordinate system. The position of the example point, lying on the surface of a sphere with radius r, is defined by a pair of head-related angles, defining the direction seen from the centre of the listener's interaural axis which coincides with the centre of the sphere. This pair of angles consists of an azimuth angle $\varphi \in [0°, 360)$, increasing counterclockwise in the horizontal plane, and a zenith or elevation angle, represented by $\theta \in [0°, 180°]$ and $\vartheta \in [-90°, 90°]$, respectively, with $\vartheta = 90° - \theta$.

In the simplified scenario, the relationship between spherical and Cartesian coordinates is given by

$$\begin{pmatrix} r \\ \varphi \\ \theta, \vartheta \end{pmatrix} = \begin{pmatrix} \sqrt{x^2 + y^2 + z^2} \\ \arctan2(-x, -z) \\ \arctan2\big(\sqrt{x^2 + z^2}, y\big), \arctan2\big(y, \sqrt{x^2 + z^2}\big) \end{pmatrix} \quad (2.1)$$

and

$$\begin{pmatrix} x \\ y \\ z \end{pmatrix} = r \begin{pmatrix} -\sin(\varphi) \cdot \sin(\theta) \\ \cos(\theta) \\ -\cos(\varphi) \cdot \sin(\theta) \end{pmatrix} = r\boldsymbol{\theta}, \quad (2.2)$$

$$\begin{pmatrix} x \\ y \\ z \end{pmatrix} = r \begin{pmatrix} -\sin(\varphi) \cdot \cos(\vartheta) \\ \sin(\vartheta) \\ -\cos(\varphi) \cdot \cos(\vartheta) \end{pmatrix}, \quad (2.3)$$

following a directional short-hand notation for $\boldsymbol{\theta}$ (Zotter, 2009a). Depending on the egocentric sound source direction, the ears at the same and opposite sides of the sound source direction are referred to as ipsilateral and contralateral ears, respectively.

According to OpenGL conventions (see, e.g. Sellers, Wright Jr, & Haemel, 2013), the listener is oriented in the negative z' direction by default, which corresponds to the conventions of the optical motion tracking system, cf. Pausch (2022, Sec. 3.1), and the real-time auralisation framework VIRTUAL ACOUSTICS (2021). The orientation of this rigid body relies on a sequence of elemental extrinsic or intrinsic rotations. Extrinsic sequences of rotation are defined around the axes of the fixed global coordinate system, whereas intrinsic ones around the current local axes assigned to the rigid body. Throughout this thesis, intrinsic rotations are used and denoted as roll, pitch and yaw angles, commonly referred to as Tait–Bryan angles (Markley & Crassidis, 2014), cf. Figure 2.1b. Each elemental rotation is sequentially performed relative to the previous rotation in the current local coordinate system of the rigid body. Note that the roll, pitch and yaw angles increase clockwise around the viewing directions in the negative z-axis, matching the default viewing direction, positive x-axis and positive y-axis, respectively. A maximum number of three sequential intrinsic rotations is sufficient to represent any rigid body orientation. Although such rotation sequences can be implemented via rotation matrices, the concept of quaternions (see, e.g. Diebel,

2006) represents a numerically stable domain and resolves the problem of gimble lock singularities (Goldstein, Poole, & Safko, 2002). Alternatively, orientations can be described by a pair of orthogonal unit-length view and up vectors, associated with the rigid body. In Figure 2.1a, the negative z'-axis and the positive y'-axis direction vectors, i.e. $[0 \quad 0 \quad -1]^{\mathsf{T}}$ and $[0 \quad 1 \quad 0]^{\mathsf{T}}$, respectively, describe the depicted nominal listener orientation via view and up vectors.

In case of any translational or rotational head movements, the local coordinate system is translated and rotated likewise. If the global and local coordinate systems do not intersect each other, which is the typical case in interactive scenarios, the relative egocentric direction depend on the current pose and position of the listener and the point, respectively. Such relationships are typically described by homogeneous transformations (see, e.g. Spong, Hutchinson, Vidyasagar, et al., 2006).

2.2 Digital signal processing

2.2.1 Discrete-time signals

In discrete-time signal processing, a continuous-time signal $x(t)$ can be represented as a sequence of numbers, sampled at equidistant time instances $nT_{\mathrm{s}} = n/f_{\mathrm{s}}$, $n \in \mathbb{N}$, determined by a sufficiently high sampling rate f_{s}, see Section 2.2.3. This sequence of numbers is expressed by $x(n)$, consisting of the values x_n (Oppenheim, Schafer, & Buck, 1999).

2.2.2 Linear time-invariant systems

Being among the most important classes of discrete-time systems, an LTI system can be uniquely and completely described by its IR $h(n)$ of length $\mathrm{M_h}$, transforming an input signal $x(n)$ of length $\mathrm{M_x}$ into an output signal $y(n)$, see Figure 2.2. Mathematically, this relationship is represented by a linear convolution (Oppenheim, Schafer, & Buck, 1999; Wefers, 2015), yielding

$$y(n) = x(n) * h(n) = \sum_{l=0}^{\mathrm{M_h}-1} x(l) \cdot h(n-l), \qquad (2.4)$$

with l denoting a discrete-time shifting index, resulting in an output signal of length $(\mathrm{M_h} + \mathrm{M_x} - 1)$. In theory, exciting an LTI system by the unit impulse

$$\delta(n) = \begin{cases} 1 & \text{for } n = 0 \\ 0 & \text{for } n \neq 0, \end{cases} \qquad (2.5)$$

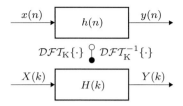

Figure 2.2: Schematic depiction of an LTI system, represented by its IR $h(n)$ in discrete-time domain, with finite time-discrete input signal $x(n)$ and output signal $y(n)$. The signals are equivalently represented in frequency domain by the transfer function $H(k)$ and the input and output spectra $X(k)$ and $Y(k)$, respectively. All corresponding signals are interconnected in time and frequency domain, and vice versa, via the forward and inverse K-point discrete Fourier transform (DFT), respectively.

allows to directly obtain its IR

$$y(n) = \delta(n) * h(n) = h(n). \tag{2.6}$$

LTI systems fulfil the properties of additivity and scaling, collectively referred to as superposition property. Given two arbitrary constants applied to two input signals will result in the corresponding linear combination at the output of the LTI system, weighted by the same constants. Additionally, the property of time invariance will maintain a time shift, applied to an input signal, in the corresponding output signal (Oppenheim, Schafer, & Buck, 1999).

2.2.3 Discrete Fourier transform

A K-point DFT allows lossless representation of a finite time-discrete signal of length M, given K = M, by M bins equidistantly spaced at f_s/M in the frequency domain (Oppenheim, Schafer, & Buck, 1999). To prevent aliasing artefacts, the frequency range of the signal must be limited to $f_{max} < f_s/2$ prior to applying the DFT transform, to fulfil the Whittaker–Shannon–Kotelnikov theorem (Whittaker, 1915; Shannon, 1948; Kotel'nikov, 1933). After zero-padding all signals in Equation (2.4) to length K = $M_h + M_x - 1$, the application of a K-point DFT to each of the signals and a subsequent inverse K-point DFT on the resulting element-wise product spectrum, denoted by $\circ$, results in (Agarwal & Burrus, 1974; Wefers, 2015)

$$\mathring{y}(n) = \mathcal{DFT}_K^{-1}\{\mathcal{DFT}_K\{x(n)\} \circ \mathcal{DFT}_K\{h(n)\}\}. \tag{2.7}$$

This relationship is known as the cyclic convolution property and yields exactly the same result as the linear convolution, i.e. $\mathring{y}(n) \equiv y(n)$ (Oppenheim, Schafer,

& Buck, 1999). If the zero-padded length of the involved signals is longer the first $(M_h + M_x - 1)$ samples of $\mathring{y}(n)$ need to be extracted. The cyclic convolution property in combination with equal signal lengths, preferably power-of-two lengths, are extensively used in classic fast Fourier transform algorithms (Vetterli & Nussbaumer, 1984) and further optimised using partitioned convolution schemes (Wefers, 2015), which are particularly relevant for applications involving real-time auralisation based on finite impulse response (FIR) filtering.

Since the involved discrete-time signals and filters in Equation (2.4) are usually real-valued in acoustic signal processing, the resulting K-point DFT spectrum is complex-conjugate symmetric by $X(k) = \overline{X(M-k)}$, containing all required information in the first $K/2 + 1$ and $(K + 1)/2$ frequency bins for even and odd (zero-padded) signal lengths, respectively (Oppenheim, Schafer, & Buck, 1999). This data redundancy allows to substantially speed up fast Fourier transform algorithms (Wefers, 2015).

2.2.4 Magnitude and phase spectra

Acoustic signal analyis often relies on the decomposition of the complex-valued DFT spectrum of a real-valued IR into its magnitude and phase spectra, yielding

$$H(k) = |H(k)| \cdot e^{j \arg\{H(k)\}}, \tag{2.8}$$

$$\text{with} \quad |H(k)| = \sqrt{\Re\{H(k)\}^2 + \Im\{H(k)\}^2},$$

$$\arg\{H(k)\} = \arctan\left(\frac{\Im\{H(k)\}}{\Re\{H(k)\}}\right),$$

$j = \sqrt{-1}$ representing the imaginary unit, and $\Re\{\cdot\}$ and $\Im\{\cdot\}$ denoting the real and imaginary part, respectively.

2.2.5 Finite impulse response filters

Discrete FIR filters represent finite IRs. Owing to their feed-forward filter structure, such filters are stable and possess a stable inverse. They only allow to process a finite number of input samples with bounded values at a time instance, producing a weighted sum with bounded result at this time instance depending on the filter coefficients, also called taps. FIR direct form structures represent efficient implementations, substantially reducing the number of required delay elements and the filter complexity. If the FIR taps are symmetric or anti symmetric the filter will exhibit a linear phase, imprinting a constant time shift on the input signal for all frequencies (Oppenheim, Schafer, & Buck, 1999). FIR filters are extensively used for applications with high demands on perceptual plausibility and phase fidelity like, for example, loudspeaker equalisation (EQ) filters.

2.2.6 Minimum-phase systems

The minimum-phase system equivalent $H_{\min}(k)$ of an LTI system $H(k)$, exhibiting an identical magnitude spectrum, can be calculated using the Hilbert transform. Minimum-phase systems are stable and causal, possess a stable and causal inverse $H_{\min}^{-1}(k)$, minimum group delay, and have all poles and zeros within or on the unit circle in the discrete frequency domain (z-domain). In the time domain, their causal IR onsets and the associated energy are minimally delayed and shifted towards $k = 0$ (Oppenheim, Schafer, & Buck, 1999). For example, in acoustics, such systems are used for decomposing head-related transfer functions (HRTFs) into their minimum-phase and all-pass components, enabling auditory modelling (Plogsties et al., 2000), or perceptually driven headphone EQ routines that require minimal phase distortion (Masiero & Fels, 2011).

2.3 Room acoustic measurements

2.3.1 Practical determination of the room impulse response

Rooms are typically considered LTI systems (Kuttruff, 2016; Vorländer, 2020), although this assumptions may be violated under time-variant environmental conditions (Guski, 2015). In practical measurement applications, $\delta(n)$ in Equation (2.5) can be approximated using impulsive acoustic sources (e.g. balloons, shotguns, starting flaps), which, however, is not recommended due to the difficult reproducibility and a poor SNR. If the room shall be excited via an omnidirectional loudspeaker (ISO 3382-2, 2008), another opposing rationale is related to the limited impulse fidelity of electroacoustic systems, which would cause large non-linearities (Klein, 2020) and potential damage due to the time-limited character of $\delta(n)$ and the associated short-term energy excitation.

Due to its superior properties, particularly in terms of achievable SNR and the possibility to elegantly reduce the influence of non-linearities in IR measurements (Farina, 2000; Müller & Massarani, 2001; Klein, 2020), exponentially swept sines $s(n)$ with different lengths are used throughout this thesis. Replacing $\delta(n)$ by $s(n)$ in Equation (2.5) allows to efficiently determine the cyclic IR via element-wise complex-valued frequency division and subsequent application of an inverse K-point DFT, yielding

$$\mathring{h}(n) = \mathcal{DFT}_{\mathrm{K}}^{-1}\left\{\frac{Y(k)}{S(k)}\right\}. \tag{2.9}$$

In the following, the cyclic IR is equated with the linear IR and used equivalently, i.e. $\mathring{h}(n) \hat{=} h(n)$. Further improvement of the SNR in the measured IRs can

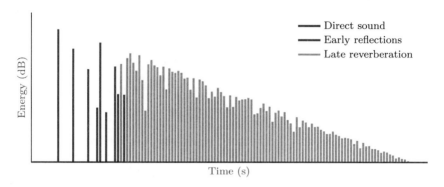

Figure 2.3: Schematic depiction of a room IR in time domain with a perception-driven separation of its components.

be achieved by either increasing the length of $s(n)$ or by calculating the IR as arithmetical average of several repeated measurement results. The latter strategy theoretically leads to an enhancement of 3 dB per doubling of measurement cycles, owing to the coherent and incoherent signal properties of $s(n)$ and $y(n)$, respectively. Application of a practically more feasible regularised spectral division is also beneficial (Farina, 2007). The frequency-dependent Tikhonov parameter is best set in a way to obtain an unaffected complex-valued element-wise spectral division in the frequency band of interest, typically between 20 Hz and 20 kHz, and suppress any measurement noise in the excluded frequency regions below and above (Sanches Masiero, 2012).

An example IR of a room is shown in Figure 2.3, with its temporal evolution separating three perceptually relevant reflection components: the direct sound (DS), early reflections and late reverberation. The DS geometrically represents the line of sight between the loudspeaker and the microphone, at the positions the IR was recorded, and is primarily important to perceptually determine the location of the sound source, cf. precedence effect (Litovsky et al., 1999). In addition to the DS, distinct reflections within 50–80 ms after the DS, reflected off room surfaces, increase the perceived loudness and apparent source width of a sound source, while also enhancing the intelligibility and clarity of speech and music, respectively. Reflections that emerge after the aforementioned time intervals support reverberation and the corresponding feeling of envelopment (Vorländer, 2020; Kuttruff, 2016)

In certain situations, for example, if it is desired that the IR only contains DS but no reflections off surfaces emerging from the measurement setup or reproduction environment, time-windowing is applied during post-processing, also ensuring finite-length IRs. This requires to perform element-wise multiplication of a window function, which is zero-valued in its stop band region, on $h(n)$. Such windows can be defined as the most basic rectangular window or using more advanced versions, such as (von) Hann, Kaiser, Dolph-Chebyshev or Blackman-Harris window functions (Prabhu, 2014). Each window function differs in terms of main lobe widths and side lobe magnitude levels, thus determining the spectral leakage pattern (Oppenheim, Schafer, & Buck, 1999). Representing a good compromise, the left and right symmetric halves of the Hann window function with different window lengths are used throughout this thesis to extract the time segment of interest from a measured IRs.

2.3.2 Room acoustic parameters

The acoustic properties of a room can be estimated on the basis of evaluating a set of IR measurements, measured for different source and receiver combinations. Based on this set of IRs, standardised room acoustic parameters can be derived. ISO 3382-1 (2009) and ISO 3382-2 (2008) cover all definitions and normative requirements for room acoustic measurements regarding the measurement equipment, measurement positions, procedural aspects, the evaluation of the IRs and the contents of the measurement report. All room acoustic measurements conducted in this thesis were performed by reference to these standards at different measurement precision levels, as stated. The hardware used for the respective measurements is described in the corresponding sections. For a detailed description of the basic elements of the generalised input/output measurement chain, the reader is referred to Klein (2020). The selected room acoustics parameters introduced below are, for example, used to characterise the experimental environments described in Pausch (2022).

Reverberation time

One of the most prominent room acoustic parameters is the reverberation time (RT). It is defined as the elapsed time in which the steady-state level, emitted by a sound source in a room, decreases by $60\,\mathrm{dB}$ after switch off (Kuttruff, 2016; Vorländer, 2020). The RT can be estimated based on the energy decay curve (EDC), defined as (Schroeder, 1965)

$$E_{\mathrm{S}}(t) = \int_{t}^{\infty} h^2(\tau)\mathrm{d}\tau = \int_{\infty}^{t} h^2(\tau)\mathrm{d}(-\tau), \qquad (2.10)$$

often referred to as Schroeder's backward integration method. With the aim of minimising the influence of the measurement noise and increasing the accuracy, the upper infinite integral limit is replaced by the point in time where the EDC transitions into the noise floor. Various compensation methods can be additionally applied to address effects related to stationary noise (Guski, 2015). The RT is subsequently derived from the normalised logarithmic EDC by linear regression analysis. Due to the limited SNR in room IRs, only the measurement data in a dynamic range between, for example, $-5\,\text{dB}$ and $-35\,\text{dB}$ is used to estimate the level decay time to $-60\,\text{dB}$, denoted by T_{30}. An estimation using other dynamic ranges works in the same way, except for the early decay time, which is evaluated between $0\,\text{dB}$ and $-10\,\text{dB}$ and experiences greater influence by the DS. However, the early decay time represents a better correlate to perceived RTs, exhibiting just noticeable differences (JNDs) of relative $5\,\%$ (ISO 3382-1, 2009). ISO 3382-2 (2008) specifies a required SNR of at least $10\,\text{dB}$, calculated at the time of the lower dynamic evaluation limit with respect to the stationary measurement noise. The typical measurement range typically covers octave bands with centre frequencies between $125\,\text{Hz}$ and $8\,\text{kHz}$. To quantify the RT of a room by a single value, the arithmetic mean of T_{30} results in the octave bands of $500\,\text{Hz}$ and $1\,\text{kHz}$, denoted by $T_{30,\text{mid}}$, may be provided for single-value characterisation (ISO 3382-2, 2008).

Clarity index for speech

When acoustic environments are required to exhibit a high level of speech intelligibility, the analysis of the temporal energy structure of room IRs is insightful. There is perceptual evidence that early reflections within the first $50\,\text{ms}$ primarily increase the clarity of a speech source in a room, thus supporting the DS, while late reflections have detrimental effects on speech intelligibility (Thiele, 1953; Seraphim, 1961). These findings motivated the energy ratio for definition (Thiele, 1953, "Deutlichkeit"), defined as

$$D_{50} = \frac{\int_0^{50\,\text{ms}} h^2(t)\,\text{d}t}{\int_0^\infty h^2(t)\,\text{d}t}, \tag{2.11}$$

with a typical range of 0.3 to 0.7 and a JND of 0.05 (ISO 3382-1, 2009). D_{50} can also be expressed as a clarity index for speech, yielding

$$C_{50} = 10\log\left(\frac{D_{50}}{1 - D_{50}}\right) \tag{2.12}$$

in units of dB. Marshall (1994) introduced a speech-intelligibility-weighted composite value C_{50}', weighting the results of C_{50} in the octave bands between $500\,\text{Hz}$ and $4\,\text{kHz}$ by 0.15, 0.25, 0.35 and 0.25, and summing up the weighted results.

Clarity index for music

Following similar considerations as for D_{50} but with a higher upper temporal limit of 80 ms, the clarity index for music is defined as (Reichardt, Alim, & Schmidt, 1974)

$$C_{80} = 10 \log \left(\frac{\int_0^{80\,\text{ms}} h^2(t)\,\mathrm{d}t}{\int_{80\,\text{ms}}^{\infty} h^2(t)\,\mathrm{d}t} \right), \tag{2.13}$$

in units of dB typically ranging between $-5\,\text{dB}$ and $5\,\text{dB}$ and a JND of $1\,\text{dB}$. The composite value C'_{80} (Marshall, 1994) represents the arithmetic mean of C_{80} values obtained for the octave bands of $500\,\text{Hz}$, $1\,\text{kHz}$ and $2\,\text{kHz}$.

Ambient noise level

Frequency-dependent ambient noise levels (ANLs), caused by external noise sources outside the laboratory and internal noise from equipment, can be considered an influential factor. The resulting noise floor is likely to affect not only the SNR in measurements but also the results of perceptual experiments. For these reasons, complementary ANL measurements were performed to further characterise the experimental environments. The results of these measurements allow to investigate the general suitability for critical listening experiments and compliance with rigorous maximum permissible ambient noise levels (MPANLs), specified by audiometry standards (ISO 8253-1, 2010; ISO 8253-2, 2009; ISO 8253-3, 2012). The calculation of ANLs does not require measured IRs but relies on the evaluation of measured time-weighted sound pressure levels (SPLs) using calibrated low-noise measurement equipment. Calibration of the relevant input measurement chain elements, i.e. microphone, microphone preamplifier and analog-to-digital (A/D) converter, was performed using a pistonphone (Type 4231, Brüel & Kjær, Nærum, Denmark), supplying an SPL of $94\,\text{dB}$ at $f = 1\,\text{kHz}$, and an external voltage source, generating a defined voltage of $1\,\text{Vrms}$ at $f = 1\,\text{kHz}$, respectively. The measured ANLs are either reported as maximum SPLs with flat (Z) frequency weighting and slow (S) time weighting, denoted by $L_{Z,S,\text{max}}$, or as equivalent continuous SPLs with flat (Z) frequency weighting, denoted by $L_{Z,\text{eq}}$ (DIN 45641:1990-06, 1990).

2.4 Source directivity

Sound signals emitted by the human mouth (Halkosaari, Vaalgamaa, & Karjalainen, 2005), musical instruments (Shabtai et al., 2017), or loudspeakers (Evans et al., 2009), are not radiated omnidirectionally but exhibit frequency-dependent and direction-dependent sound radiation patterns. Typically, such directivity patterns are measured in hemi-anechoic or anechoic chambers, cf. Pausch (2022, Sec. 2.4 and 2.5), in the far field, where the dimension of the source is much smaller than the measurement radius r. The source directivity at a specific point (r, φ, ϑ) in the far field relative is defined as (Evans et al., 2009)

$$D_0(k, r, \varphi, \vartheta) = \frac{P(k, r, \varphi, \vartheta)}{P_0(k, r, \varphi_0, \vartheta_0)}, \tag{2.14}$$

representing the ratio of the complex sound pressure spectrum measured at this point normalised to the sound pressure spectrum measured at the on-axis reference point $(r, \varphi_0, \vartheta_0)$. Special measurement setups, cf. Pausch (2022, Sec. 3.8), allow to specify a dense spatial grid of measurement directions for an accurate characterisation of source directivity. This grid typically includes directions on the horizontal and vertical measurement planes, as well as the on-axis direction of the device under test. In most instances, the magnitude values of $D_0(k, r, \varphi, \vartheta)$ in dB are visualised in two-dimensional or three-dimensional plots, the latter optionally with corresponding colour-coded phase.

2.5 Signal processing toolbox

All IR measurements, the calculation of room acoustic parameters, and most offline signal processing in this thesis were performed using the ITA Toolbox (Berzborn et al., 2017). This open-source framework of M A T L A B™ classes and routines for acoustic measurements and signal processing has been developed in-house over the last decades and can be used in combination with various measurement hardware. The measured systems were assumed to be ideally linear and time invariant. As excitation signal, exponentially swept sines were applied with different signal durations, covering a frequency range of 20 Hz–20 kHz. With the aim of increased efficiency and reduced measurement time, the multiple exponential sweep method (Majdak, Balazs, & Laback, 2007), further optimised by Dietrich, Masiero, and Vorländer (2013), was applied in selected measurements, see also (Sanches Masiero, 2012; Klein, 2020). If not stated otherwise all measurements were performed at a sampling rate of $f_\mathrm{s} = 44.1\,\mathrm{kHz}$.

2.6 The human auditory system

2.6.1 Peripheral auditory system

The human peripheral auditory system, see Figure 2.4, includes the outer ear, the middle ear, the inner ear, and the auditory nerve, which enables signal trans-

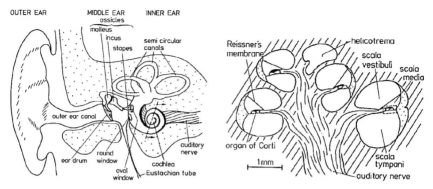

(a) Cross section of the outer, middle and inner ear.
(b) Cross section of the inner ear.

Figure 2.4: Schematic depiction of the peripheral human auditory system. (Figures taken from Zwicker and Fastl, 2013.)

mission to the auditory cortex to perceive a sound event (see, e.g. Moore, 2012; Zwicker & Fastl, 2013).

In the outer ear, the air-borne oscillations of an incident sound wave travel through the ear canal and stimulate the ear drum to vibrate. The ear canal can be approximated by a $\lambda/4$-type resonator with a resonance frequency f_0 between about 3 and 4 kHz, assuming individual canal lengths between 2.1 and 2.9 cm.

These vibrations are converted into structure-bourne sound by the three auditory ossicles of the middle ear, that is the malleus, incus and stapes, and transmitted via the oval window to the cochlea, filled with perilymphatic fluid. The different lengths of malleus and incus in combination with the surface ratio between the ear drum and the oval window aim to adjust the mismatch between the smaller specific acoustic impedances of air (about $414\,\mathrm{kg\,m^{-2}\,s^{-1}}$ at $20\,^\circ$C) and the fluid (comparable to the one of water with $1.48\mathrm{e}6\,\mathrm{kg\,m^{-2}\,s^{-1}}$ at $20\,^\circ$C) (Vorländer, 2020). The round window oscillates in opposite phase to the oval window, resulting in a pressure difference and a movement of the perilymphatic fluid (Moore, 2012). Apart from serving as an impedance transformer, specific muscles help to protect the inner ear from too loud sounds, dampening the mechanical movement of the stapes and thus restricting sound transmission particularly in the lower frequency range. This muscle contraction is commonly referred to as stapedius reflex, which is anticipatorily activated by the own voice about 100 ms before speaking, and after approximately 10 ms and 100 ms (maximum tension) subject to external sounds with SPLs in the range of 70 dB to 85 dB (Lidén, Nordlund, & Hawkins, 1964; Niemeyer & Sesterhenn, 1974). In addition to the described sound transmission path, substantially lower portions of vibrations are transmitted via the cranial bones to directly stimulate the cochlea.

The main anatomical elements of the inner ear include the spiral-shaped cochlea with bony walls, ending in the helicotrema. The vestibular system, consisting of three semi-circular canals, that is, the scala vestibuli, separated from the scala media by the Reissner's membrane, and the scala tympani, represents the equilibrium organ. The basilar membrane, see dashed rectangular window in Figure 2.4b, separates the scala media from the scala tympani. The mechanical oscillations from the oval window result in a travelling wave with a frequency-dependent peak at a specific position on the basilar membrane. This physiological phenomenon is called tonotopy and results from the relative motion between basilar and tectorial membranes. As one of the its main functions, the organ of Corti, situated on the basilar membrane, transforms these oscillations of the travelling wave into electrochemical pulses which are subsequently transmitted to the auditory cortex via the the auditory nerve. While the inner hair cells effectively produce the nerve impulses, the outer hair cells serve as mechanical amplifier, adaptively adjusting the dynamic range of the input signal. Finally, the combined auditory information of the cochlea and the vestibular organ are processed in the auditory cortex. Finally, the interaction of a complex network of neuronal connections and centres stimulates a sound sensation (Moore, 2012).

2.6.2 Measuring the auditory threshold in quiet

The human auditory system is able to perceive frequencies between $20\,\text{Hz}$ and $20\,\text{kHz}$, the upper limit gradually decreasing with age. SPLs between about $-6\,\text{dB}$ (at the ear canal resonance) up to 120–$130\,\text{dB}$ (threshold of pain) can be resolved with varying dynamic range, see normal equal-loudness-level contours (ISO 226:2003, 2003). A comparison of frequency-dependent auditory thresholds in quiet with the average results obtained from a large sample of young participants with NH allows to check the individual hearing status.

ISO 3382-1 (2009) specifies the requirements and procedure to perform subjective airborne and bone-conduction PTA. Especially the conduction of airborne PTA is considered a standard clinical test before listening experiments to ensure that the tested participants meet the pre-defined inclusion criteria. In such tests, the participants are typically presented a pulsating (amplitude-modulated) sine tone, whose level starts below the auditory threshold, via highly-insulating circumaural headphones (e.g. HDA 300, Sennheiser, Wedemark, Germany). The test tone level is subsequently increased in a stepwise manner and the participants are asked to provide feedback as soon as it is perceived, the threshold being recorded by the examiner. This ascending tone audiometry typically starts with a test tone at $f = 1\,\mathrm{kHz}$ and is repeated for octave band centre frequencies between $125\,\mathrm{Hz}$ and $8\,\mathrm{kHz}$, with additional test frequencies at $3\,\mathrm{kHz}$ and $6\,\mathrm{kHz}$. Alternatively, constant sine tones, frequency-modulated sine tones (wobble tones) or narrowband random noise with spectral bandwidths of octaves or third-octaves around the corresponding centre frequencies may be used as test stimuli (ISO 8253-2, 2009). Other test methods include, for example, the descending audiometry, the bracketing method (Arlinger, 1979), or the Békésy audiometry (Von Békésy, 1947). For complementary assessment, a bone-conduction audiometry can be performed. In this examination, the stimuli are reproduced via a bone-conduction transducer (e.g. B71W, RadioEar, Middelfart, Denmark), which is attached to the mastoid of the measured ear, to test the hearing thresholds at the aforementioned test frequencies between $250\,\mathrm{Hz}$ and $6\,\mathrm{kHz}$.

2.6.3 Hearing loss

Types of hearing loss

Various disruptions in the anatomical auditory chain and higher-level auditory processing disorders can lead to different types of HL. These types can be classified roughly in conductive and sensorineural HL, with possible mixed variants, and neural HL combined with central auditory processing disorder (Dillon, 2012). Conductive HL roots in defects located in the outer or middle ear and may be caused, for example, by objects blocking the ear canal, inflammation of the ear canal or middle ear, tumors in the respective parts, congenital or caused deformations of the ear canal, a damaged eardrum, as well as otosclerosis or defective connection of the ossicles. Each of these causes hinder proper transmission of the mechanical oscillations to the inner ear. Sensorineural HL can be connected with problems in the inner ear, like presbycusis (age-related HL), effects of (inherited) diseases or viruses, as well as anatomical malformations of the inner ear or, among the most common reasons, regular exposure to excessively loud noise, referred to as noise-induced HL. The latter may prominently affect functionality of the inner or outer hair cells and their connected synapses, including the risk of developing dead regions of hair cells on the basilar membrane. The lacking ability of the auditory nerve to pass on neuronal information to the brain or problems when processing sounds in higher auditory levels are typically connected with a neural HL or central auditory processing disorder. Spatial processing disorder is considered a specific type of auditory processing disorder and describes the limited ability to separate a spatialised target source in the presence of competing noise sources (Glyde et al., 2013), the extent of which can be measured by application of binaural SiN tests, such as the Listening in Spatialized Noise–Sentences (LiSN-

S) test (Cameron, Dillon, & Newall, 2006). This test is applied in the clinical
experiment in Chapter 6.

Degree of hearing loss

Pure-tone audiograms allow to classify the type and extent of HL as well as
its manifestation (unilateral/bilateral, asymmetric/symmetric) by means of normalised representation of the individual hearing thresholds. Figure 2.5 displays
an example audiogram, including air conduction and bone conduction measurements. While the left-ear results indicate sensorineural HL with both transmission
paths being similarly disrupted, the right-ear results hint some sort of disruption in the outer or middle ear with normal inner ear functionality, suggesting
conductive HL. Classification schemes, such as the one presented by Clark et
al. (1981), are used to categorise the degree of HL (Baiduc et al., 2013), see
Figure 2.6. Specifically, the four-frequency pure tone average (4-PTA), i.e. the
average of audiometric results for the test frequencies 500 Hz, 1 kHz, 2 kHz, and
3 kHz (Gurgel et al., 2012), and the resulting slope of the corresponding linear
regression line can be provided. Further objective clinical examinations include
tympanometry (Lidén, Peterson, & Björkman, 1970), a method allowing to assess
the condition of the ear drum and the middle ear, as well as measurements of
otoacoustic emissions (Probst, Lonsbury-Martin, & Martin, 1991) and auditory
evoked potentials (Hall III, 2007).

Effects of hearing loss on (spatial) hearing

Dillon (2012) provides various factors to explain perceptual effects related to
sensorineural HL on general sound perception, particularly on speech, and, in
part, spatial hearing, that are summarised briefly below.

Reduced audibility.   Malfunctioning or dead regions in the inner hair cells are
one of the main factors to prevent proper frequency-dependent transmission of
electrochemical pulses along the auditory nerve. The degree of HL determines
the amount of information loss, for example, by hindering unambiguous classification of consonants in speech, which is vital to grasp individual words and follow
continuous conversations. Closely related is the issue that sensorineural HL with
high-frequency sloping typically impedes the perception of formants, i.e. resonances of the vocal tract resulting in a spectral energy accumulation, that are
necessary to identify vowels. Accounting for the fact that speech tends to have
a decaying spectral energy distribution towards higher frequencies and HL often desensitises the frequency range between 500 Hz and 4 kHz and above, it is
likely that affected individuals perceive speech as muffled and struggle to under-

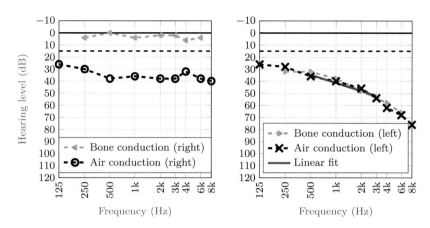

Figure 2.5: Air-conduction and bone-conduction PTAs of an individual with sensorineural and conductive HL in the left and right ear, respectively. The left air-conduction audiogram exhibits a 4-PTA of 44 dB and a slope of −6.7 dB/octave, represented by the red linear regression line. The solid and dashed horizontal lines indicate the average auditory threshold in quiet of individuals with NH and the limit for NH classification as per Clark et al. (1981), respectively.

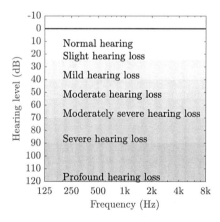

Figure 2.6: Classification of the degree of HL as per Clark et al. (1981).

stand and process the contents, which is particularly challenging in multi-talker situations.

Reduced dynamic range.   The increase of the auditory threshold in individuals with HL consequently results in a decreased accessible dynamic range. However, the loudness discomfort level does not increase to the same extent, resulting in a steeper loudness increase, a phenomenon referred to as recruitment (Zwicker & Fastl, 2013). Recruitment is primarily related to some disruption of the outer hair cells and their main function of dynamic range control or an otherwise impaired cochlear system. Basically, the cochlear amplifier represents a frequency-selective non-linear amplifier, adaptively reacting to the SPL of input sounds, explained in more detail in the following paragraph.

Reduced frequency resolution.   The frequency-selective structure of the basilar membrane (Von Békésy & Wever, 1960) allows to discern the spectral contents of an input signal by analysing the spatial resonance pattern of the resulting travelling wave. While higher frequencies excite the basilar membrane close to the oval window, lower frequencies result in resonances closer to the apex, which is favoured by a narrower and stiffer, and wider and looser geometry of the basilar membrane, respectively (Zwicker & Fastl, 2013). Moore (2012) provides a review of techniques to measure frequency tuning curves, the results of which suggest that their sharpness depends on the physiological condition of the tested animals. The degree of sharpness of the tuning curves is determined by the cochlear amplifier, whose active non-linear functioning and the corresponding frequency selectivity worsens as shown via experiments involving animals with gradually decreasing physiological condition (Sellick, Patuzzi, & Johnstone, 1982). These results are transferable to individuals with HL, who will likely suffer from the same effects due to a combination of damaged inner and outer hair cells and an accompanying limited functionality of the cochlear amplifier.

Closely related, simultaneous spectral masking effects are reflected by classical masking patterns (Zwicker & Fastl, 2013). The masking patterns were obtained in presence of narrowband noise with a specific centre frequency while varying the frequency and level of a test tone so that it is masked by the noise. While the resulting patterns exhibit steeper slopes towards lower frequencies, flatter slopes, gradually decreasing with increasing masker level, are observed towards higher frequencies. These results can be linked to the aforementioned frequency-dependent excitation regions on the basilar membrane in combination with the direction of the travelling wave, causing the so-called upward spread of masking (Moore, 2012), which further hinders to perceive high-frequency consonants when additionally suffering from negatively sloping sensorineural HL.

Transferred to speech comprehension, these observations allow to explain the detrimental effect that a target and interference signals do not result in two separated excitation regions on the basilar membrane but rather produce a single broadened resonance, making it challenging or even impossible to segregate the spectral contents of the target speech.

Reduced temporal resolution. Masking effects also occur in time domain and can be separated in pre-masking, simultaneous masking and post-masking stages. The related experiments rely on a temporarily varied presentation of test-tone bursts or short pulses that are shifted relatively to a masker, for example, a 200-ms noise train. The effect of pre-masking, or backward masking, is still not well understood (Zwicker & Fastl, 2013; Moore, 2012) and occurs in a time range of 20 ms before the masker. According to Zwicker and Fastl (2013), this effect is not related to acausal mechanisms of our hearing system but to the slower and faster build-up times for the test and masker signals, respectively. In comparison, post-masking, or forward masking can be easier explained by accounting for a necessary recovery time of the hair cells after short-term stimulation, requiring between 100 and 200 ms (Zwicker & Fastl, 2013). The effect shows only little dependence on the level of the masker, but rather on its duration (Moore, 2012). Simultaneous temporal masking in the presence of a white-noise masker depends on the duration and repetition rate of a test-tone burst in relation to the SPL of the masking noise. Since the masked thresholds increase for test-tone burst durations smaller than 200 ms, or corresponding repetition rates of 5 Hz, it is assumed that our hearing system applies the same integration interval (Zwicker & Fastl, 2013).

These temporal masking phenomena are more prominent in individuals with HL and impair speech intelligibility in real-life conversational settings. Given two competing fluctuating speech and background noise signals, elaborate listening strategies, such as listening in the gaps (temporal glimpsing) or exploiting the temporal fine structure of speech signals (Dillon, 2012), are unconsciously applied to extract all tangible fragments of information from the continuous target speech stream. Such listening strategies cannot be applied with the same efficiency by individuals with deteriorated hair cells, resulting in reduced temporal precision of sending neural impulses and potentially jittered signals, collectively degrading the stream of information.

Elevated signal-to-noise ratio requirements. Elevated hearing thresholds and a limited dynamic range, decreased spectral and temporal adaption capabilities, as discussed above, reduce speech intelligibility and flexibility of the auditory system in time-varying noisy environments, generally requiring between 4 to 10 dB higher

SNR levels (Dillon, 2012). Individuals with HL, additionally suffering from spatial processing disorder (Glyde et al., 2013), typically show poorer performance in SiN experiments, represented by higher speech reception thresholds (SRTs). An introduction to the fundamental concepts on the determination of such thresholds is provided in Section 2.9.

2.7 Hearing aids

2.7.1 Types of hearing aids

HAs represent a technical intervention tool against HLs and can be categorised according to the device and housing styles, see Dillon (2012). As the name suggests, the housings of behind-the-ears (BTEs) HAs are placed behind the listener's ears. Various electroacoustic components are contained, including a set of microphones, an on-board DSP unit, a telecoil facilitating the use of public induction loop access points or telephones, a battery, and user control elements. Depending on the degree of HL, sound transmission to the ear canal either relies on a vent for acoustic signal transmission with an individual ear mould attached, or a miniature loudspeaker, commonly referred to as HA receiver, to be inserted into the outer part of the ear canal using an attached silicone ear piece. In the former BTE device variant, the receiver is integrated in the housing, whereas the externalised variant requires the receiver to be included at the end of a slim cable for electronic signal transmission, connected to the housing, thus called BTE receiver-in-the-ear canal (RITE) device. A prototype of such a device is presented in (Pausch, 2022, Sec. 3.4) and used for all experiments in this thesis. Other HA device styles shall be mentioned for the purpose of completeness and include in-the-ear HAs, completely-in-the-canal HAs, spectacle or eyeglass aids, or larger external body-worn aids (Dillon, 2012).

2.7.2 Hearing aid algorithms and their objectives

A complex network of HA signal processing algorithms aims to mitigate the detrimental effects of HL, cf. Section 2.6.3. Figure 2.7 presents a generic block diagram of a modern HA (single device). An acoustic sound signal is picked up by a set of typically two or three microphones, amplified and A/D-converted to be further processed by the on-board DSP unit. The beamforming stage exploits the spatial separation of the HA microphones and the resulting differences in the times of arrival (TOAs) to establish a direction-dependent pattern steered to the sound source of interest, for example, the target talker in a conversational scenario. Ranging from simple static unilateral delay-and-sum beamformers to more complex adaptive binaural beamforming algorithms, such as minimum-variance distortion-

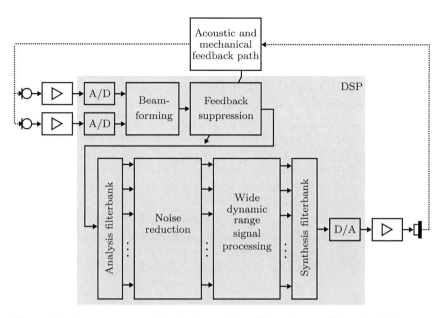

Figure 2.7: Generic overview of a dual-microphone HA with typically applied DSP algorithms. (Figure taken from Hamacher et al., 2005, and adapted.)

less response beamformers (Van Trees, 2004; Adiloğlu et al., 2015), directional
signal processing allows to provide substantially improved SNRs and corresponding speech intelligibility (see, e.g. Picou & Ricketts, 2019; Best et al., 2015). The
close arrangement of the HA microphones and the HA receiver, particularly in
(completely-)in-the-canal HAs represent a potential source of mechanical and
acoustic feedback, indicated by the dashed feedback path, which determines the
current parameter setting of adaptive feedback cancellation algorithms (Kates,
2003). The bounded output signal of the feedback suppression stage is subsequently split into a number of frequency bands, using an analysis filterbank with
channel bandwidths that ideally match the frequency resolution of the basilar
membrane, characterised by critical bands (Zwicker & Fastl, 2013), and a corresponding synthesis filterbank (see, e.g. Hohmann, 2002). This approach allows to
increase the efficiency of the subsequent noise reduction stage (Bentler & Chiou,
2006) by tuning the parameter sets per frequency band. As obvious as the goal
of reducing noise may sound in theory, the classification of the noise component
in a complex signal not only highly depends on the environment and its context
but also on user preferences. The application of scene classification algorithms,
relying, for example, on blind estimation methods of sound field parameters or

deep learning (Scharrer, 2014; Vivek, Vidhya, & Madhanmohan, 2020), provide powerful approaches to further improve the challenging task of automatically selecting the appropriate beamforming and noise reduction schemes. The afore-mentioned restricted dynamic range access and related recruitment issues can be best addressed by means of wide dynamic range signal processing algorithms. Such algorithms allow to map the dynamic range of an acoustic input signal to the available individual dynamic range, which is accounted for by the processed output signal. Since a purely linearly amplified input signal would potentially exceed the threshold of discomfort or even pain, the algorithm class of com-pressors aims to increase too faint input levels ensuring their audibility, linearly amplifies mid-range levels, and appropriately limits too loud sounds, making the best possible use of the available dynamic range (Kuk, 1996; Kates, 2010). The compressive behaviour can be controlled by defining characteristic input-output compressor curves, including the compressor thresholds and ratios per frequency band, as well as time settings that determine attack and release behaviour (Dillon, 2012). Underlying compression speed rationales include automatic gain control (slow compression), aiming to preserve the overall long-term level, or syllabic compression (fast compression) to improve speech intelligibility by a reduction of level fluctuations between syllables, in this way addressing the aforementioned limited temporal resolution due to HL. After applying the synthesis filterbank, the single-channel output signal is digital-to-analog (D/A)-converted, amplified and reproduced via the HA receiver, whose location and characteristics depend on the BTE HA device style and the fitting type, respectively.

2.7.3 Hearing aid fitting strategies

Speaking of individual settings in HA algorithms, the parameters related to the frequency-dependent wide dynamic range signal processing algorithms, including insertion gains and compressor settings, are calculated following fitting rationales. Finding an appropriate user-specific fitting and an acceptable balance between loudness perception, speech intelligibility and overall sound quality represents one of the most challenging tasks for audiologists since it requires to account for a number of patient-related factors including age, the type and degree of HL, individual ear (canal) geometry, device-specific and algorithm-specific settings, as well as personal preferences. A comprehensive overview of HA fitting strategies is provided by Palmer and Lindley (2002), and Dillon (2012).

In an audiologist-driven approach, the first-fit settings aim to achieve the pre-scriptive targets of the selected fitting routine, as calculated based on the patient's individual auditory threshold in quiet as well as dynamic range measurements including the loudness discomfort level and loudness contour data. These settings are subsequently refined in several sessions in close collaboration with the patient, accounting for possible dissatisfaction and individual preferences. However, the outcome of this approach strongly depends on the awareness of the patient to communicate these preferences. On the other hand, self-adjusted gain settings, as part of a patient-driven strategy, may produce substantial deviations from those prescribed by typically applied fitting calculation formulas (Nelson et al., 2018) and motivate the use of a combined fitting approach. Prescription settings can be verified based on 2-cc coupler measurements (IEC 60318-5:2006, 2006), representing average adult ear canal data. For increased precision, individual ear canal measurements using probe tube microphones are conduced to derive the so-called real-ear-to-coupler difference. Such measurements are particularly important for children, for whom there are currently no standardised ear couplers available, likely resulting in increased fitting errors when based on 2-cc coupler results (Fels, 2006).

Due to their superior dynamic range fitting capability, non-linear target prescription rules are usually preferred over linear fitting strategies. Popular examples of such non-linear approaches include

- FIG6 (Killion & Fikret-Pasa, 1993),
- desired sensation level (Cornelisse, Seewald, & Jamieson, 1995),
- National Acoustic Laboratories non-linear prescriptions (NAL-NL1 / NAL-NL2) (Byrne et al., 2001; Keidser et al., 2011), and the
- Cambridge fitting formula with multi-channel compression (Moore et al., 1999).

The last mentioned fitting strategy has been applied in the clinical experiment presented in Chapter 6 to create the first fit for the HA users. On the one hand, its rationales aim at restoring the loudness pattern for moderate speech levels, that is about 65 dB, and should result in a similar equal loudness within a critical band as perceived by individuals with NH. On the other hand, quiet speech around 45 dB in the frequency bands with centre frequencies between 500 Hz to 4 kHz should be amplified to the level of audibility with the constraint of limiting the by-channel compression ratios to a maximum of 2.92, considering the detrimental effects of too much compression on speech intelligibility both in quiet and in noise. Moore et al. (1999) credits the fitting strategy with good loudness perception and speech intelligibility for various acoustic signal levels and observed only slight changes in algorithmic parameter settings between the initial fit and after fine-tuning.

2.8 Combined binaural hearing using hearing aids

Figure 2.8 schematically presents how individuals fitted with BTE RITE HAs perceive a spatialised sound source, located in the distal region (Brungart & Rabinowitz, 1999). It is easy to see that the sound source will be perceived combined via the HAs and in part via the "natural" path through the blocked ear canal entrance. Representing a fundamental concept in binaural technology, the influence of the torso, shoulder, head and pinna act as a directional filter, exhibiting highly individual features in the time and frequency domain (Blauert, 1997), which will be introduced in the sections below. In the HA-related path, the filtered signal is captured by the microphones of each HA device at some location behind the respective ear and subsequently processed by the HA algorithms. This will result in a processing delay, denoted as Δt_{HA}, before the signals are played back via the HA receivers with attached silicone ear pieces, which are inserted into the entrance of the ear canals. Depending on the type and degree of HL and the passive damping properties of the ear piece, the individual may have access to residual hearing, allowing to perceive parts of the external sound field in a natural way through the blocked ear canal entrance. The final signal will consist of these summed transmission path components, travelling through the remaining ear canal to stimulate the ear drum to vibrate.

2.8.1 Head-related transfer functions

Assuming a scenario with a listener with NH, the air-borne sound oscillations, emitted by the sound source in the distal region of the egocentric three-dimensional space are filtered direction-specific, depending on their frequency and corresponding wavelength (Blauert, 1997; Møller et al., 1995b). This filtering process results in binaural and monaural cues, which are crucial for sound source localisation. According to Blauert (1997), the free-field HRTF of the left (L) and right (R) ear is defined as

$$H_{\{\text{L, R}\}}(k, \varphi, \vartheta) = \frac{H_{\text{ear, }\{\text{L, R}\}}(k, \varphi, \vartheta)}{H_{\text{ear, ref, }\{\text{L, R}\}}(k, \varphi, \vartheta)}. \qquad (2.15)$$

It represents the ratio of the transfer function between the sound source to a measurement point in the ear canal, which can be located at the blocked ear canal entrance or close to the ear drum (Oberem, 2020; Denk & Kollmeier, 2021), and the transfer function for the same source-receiver scenario with absent individual, measured at the centre of the interaural axis. This type of HRTF, measured at the blocked ear canal entrance, will be used throughout the thesis. For increased readability, the direction dependency will no longer be explicitly stated. In time

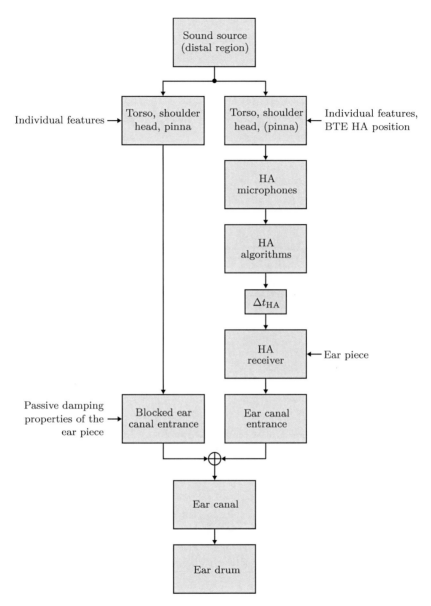

Figure 2.8: Combined binaural perception of a spatialised sound source located in the distal region by an individual fitted with BTE RITE HAs.

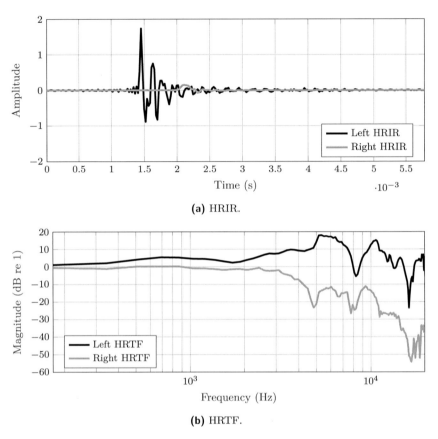

(a) HRIR.

(b) HRTF.

Figure 2.9: Example binaural receiver directivities of length M = 256 samples in time and frequency domain, measured from $(\varphi, \vartheta) = (90°, 0°)$ at the blocked ear canal entrance.

domain, HRTFs are referred to as head-related impulse responses (HRIRs) and linked via the inverse DFT. Figure 2.9 shows an example of an individual binaural receiver directivity in time and frequency domain for a direction of incidence of $(\varphi, \vartheta) = (90°, 0°)$. The influence of the respective anthropometric parts becomes visible in the HRTF magnitude spectra above about 200 Hz with increasingly detailed spectral patterns towards higher frequencies.

Interaural time differences

In time-domain, the finite propagation speed of sound waves in air results in delayed TOAs of the source signal at the two ear canal entrances, cf. Figure 2.9a, their difference referred to as interaural time difference (ITD). The ITDs exhibit a sine-shaped curve progression for source directions in the horizontal plane with minimum values for directions lying on the median plane and maximum values in the range of $(\varphi, \vartheta) = (90°, 0°)$ and $(\varphi, \vartheta) = (270°, 0°)$, if fulfilling the prerequisite of an antipodal arrangement of the ear canal entrances (Aaronson & Hartmann, 2014). Katz and Noisternig (2014) provides a methodological overview of different methods to calculate this binaural cue. Figure 2.10a shows the ITDs of the example HRTF, represented by the black solid curve. The results were obtained using the interaural cross correlation (IACC) method, which is one of the perceptually most relevant methods (Andreopoulou & Katz, 2017), for directions of incidence of $\varphi \in [0°, 360)$. A 10-th order low-pass digital Butterworth filter with low-pass filter with a cut-off frequency of 1.5 kHz was applied to remove unwanted noise. According to the Duplex theory coined by Lord Rayleigh (Rayleigh, 1907), this binaural cue is particularly relevant for horizontal source localisation for frequencies below about 1.5 kHz, above which its evaluation becomes ambiguous and misleading (Bernstein & Trahiotis, 1985; Hartmann & Macaulay, 2014).

Interaural level differences

Interaural level differences (ILDs) become the dominant binaural cue for horizontal sound source localisation above about 1.5 kHz (Blauert, 1997). As the wavelength of the incident wave decreases with increasing frequency, the head shadow effect will result in lower amplitudes at the contralateral ear compared to the ipsilateral ear. Although this binaural cue is reflected by a lowered amplitude in the HRIR of the contralateral ear, its highly frequency-dependent characteristics can be better analysed in the frequency domain, cf. Figure 2.9. The ILD is defined as

$$\text{ILD}(k, \varphi, \vartheta) = 20 \log\left(\left|\frac{H_\text{L}(k, \varphi, \vartheta)}{H_\text{R}(k, \varphi, \vartheta)}\right|\right), \tag{2.16}$$

representing the logarithmic magnitude ratio of the complex opposite-ear HRTFs for a given direction of incidence.

Monaural and dynamic localisation cues

On spatial regions with constant ITD and ILD values, so-called cones of confusion (Mills, 1972), binaural cues may not be sufficient to unambiguously localise a spatialised sound source. An example for such a region is the median plane.

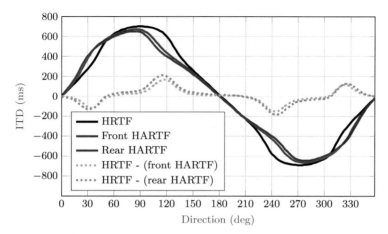

(a) ITDs estimated for directions of incidence $\varphi \in [0°, 180°]$ in the horizontal plane based on IACC in a frequency range of 500 Hz to 1.5 kHz.

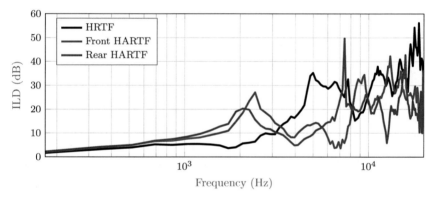

(b) ILDs for a direction of incidence of $(\varphi, \vartheta) = (90°, 0°)$, evaluated in a frequency range of 1.5 kHz to 20 kHz.

Figure 2.10: Binaural cues for the different types of an example binaural receiver directivity.

A reduction of reversals, i.e. erroneously localising a sound source in the opposite half space, can be achieved by additionally evaluating monaural cues. These cues are provided by the fine structure of the pinna and the corresponding wave interaction effects, producing distinct spectral patterns including narrowband notches in the high frequency region (Wright, Hebrank, & Wilson, 1974; Blauert, 1997). Utilising dynamic cues by changing the head orientation further helps to improve the localisation performance (Thurlow & Runge, 1967). In this context, the term binaural sluggishness refers to the ability to follow the spatial trajectory of a sound source by corresponding head movements (Culling & Colburn, 2000; Viveros Munoz, 2019).

2.8.2 Hearing aid-related transfer functions

The fundamental concept of HRTFs can be transferred to spatial sound perception via HAs. As presented in Figure 2.8, the individual anthropometric features similarly influence the sound field before being captured at the microphone positions of a HAs. The resulting direction-dependent hearing aid-related transfer functions (HARTFs) are defined analogously to the HRTF counterparts by

$$H'_{\{L, R\}, \{front, rear\}}(k, \varphi, \vartheta) = \frac{H_{HA, \{L, R\}, \{front, rear\}}(k, \varphi, \vartheta)}{H_{HA, ref, \{L, R\}, \{front, rear\}}(k, \varphi, \vartheta)}. \qquad (2.17)$$

Such spatial transfer functions are measured at the front or rear microphone positions of a dual-microphone BTE HA, just like the used research HA, see (Pausch, 2022, Sec. 3.4), while being attached to the listener. The result is divided by the transfer function between the source and the HA with absent listener at the centre of the interaural axis. As for HRTFs, the direction dependency will no longer be explicitly stated. Figure 2.11 shows the resulting directional spatial filter in time domain, called hearing aid-related impulse responses (HARIRs), and frequency domain, for the same participant and direction as in Figure 2.9. It can be seen that a small spatial displacement between the respective HA microphone pairs already leads to spectro-temporal deviations, which become particularly important for HA algorithms that exploit spatial sound field features (Hamacher et al., 2005; Dillon, 2012; Kates, 2008).

Differences in localisation cues

Although the same principal localisation cues as in HRTFs can be observed in HARTF datasets, their spectro-temporal characteristics exhibit substantial differences. With reference to Figure 2.10a, the ITDs in both front and rear HARTF deviate from the ones estimated for HRTFs. This deviation can be attributed to the spatial displacement of the HA microphones behind the listener's ear off the lateral head centre. The shortened interaural path lengths result in ITD maxima with reduced magnitudes. Additionally, the arguments of the ITD maxima already appear at smaller and greater directions of incidence in the left and right half spaces, respectively, since the resulting horizontal HA microphone angles with respect to the median plane are shifted more rearwards compared to the ear canal entrances (Aaronson & Hartmann, 2014; Pausch, Doma, & Fels, 2021).

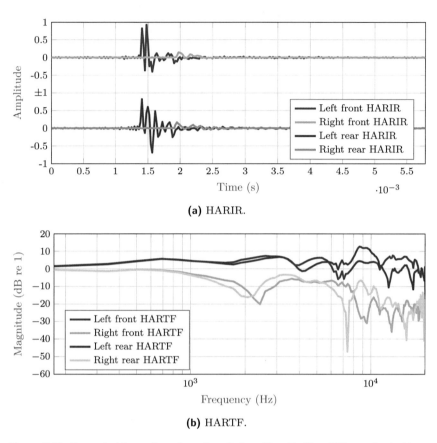

(a) HARIR.

(b) HARTF.

Figure 2.11: Example binaural receiver directivity of length $M = 256$ samples in time and frequency domain, measured from $(\varphi, \vartheta) = (90°, 0°)$ at the front and rear microphones of a BTE HA.

Differences in ILDs between HRTFs and HARTFs, as presented in Figure 2.10b, appear for frequencies above about $500\,\text{Hz}$ and become more pronounced with increasing frequency. Strong variations between the front and rear HA microphone pairs can be observed. Due to pronounced off-head centre position, the ILD magnitudes tend to be smaller than that of HRTF. However, this characteristic is highly dependent on individual anthropometric features, HA placement, the frequency and the evaluated direction of incidence (Pausch, Doma, & Fels, 2021). The HA microphone placement behind the ears substantially changes the influence of the pinna compared to the conventional listening scenario via HRTFs.

This aspect typically results in a generally lowered dynamic range in the corresponding magnitude variations and less pronounced notches on the ipsilateral ear side.

2.9 Speech-in-noise perception

In consideration of the aforementioned issues related to HL, cf. Section 2.6.3, affected individuals often have problems to perceive and comprehend speech in conversational settings in the presence of interfering noise or distractors (Dillon, 2012; Houtgast & Festen, 2008). Any disruption along the auditory pathway may also impact the ability to direct one's focus to a target source while ignoring interfering sources (Cherry, 1953; Bronkhorst, 2000). Increasingly challenging room acoustic conditions, typically correlating with higher RTs and lower clarity values for speech, see Section 2.3.2, lead to a further deterioration of speech perception performance (Culling, 2016), particularly in individuals fitted with HAs and already at a young age (Iglehart, 2020). To better quantify the impact of HL on speech perception in everyday listening situations, it is therefore crucial to complement the measurements on individual hearing thresholds in quiet, see Section 2.6.2, by SiN tests, ideally addressing binaural hearing (Bronkhorst & Plomp, 1990; Wagener, Brand, & Kollmeier, 1999; Cameron, Dillon, & Newall, 2006; Soli & Wong, 2008) and its advantages for auditory selective attention (Oberem, 2020).

2.9.1 Speech reception thresholds

The rationale of SiN tests relies on varying the SNR between a target talker and a single or multiple interfering noise source(s). This is accomplished by either changing the level of the target talker or the noise source(s), while keeping the other component fixed, to adaptively estimate the SRT based on the participant's responses. According to the definition by Plomp (1978), the SRT is defined as the SNR in dB that is required to understand 50 % of the presented target speech material – a percentage that may be varied (see, e.g. Messaoud-Galusi, Hazan, & Rosen, 2011) – representing the corresponding location on the underlying psychometric function (Wichmann & Hill, 2001). A psychometric function predicts a response variable, e.g. the proportion of correct responses in an SiN paradigm, in relation to the varied levels of a physical stimulus, e.g. the SNR levels. Figure 2.12 shows an example of such a psychometric function, with the 50 % SRT predicted at about 3 dB.

Possible methods to adaptively determine the SRT range from simple staircase procedures (Levitt, 1971), over parameter estimation by sequential test-

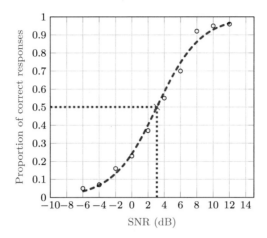

Figure 2.12: Example of a psychometric function (blue dashed line), modelling the relationship between the proportion of correct responses on the *y*-axis and the varied SNR levels on the *x*-axis (black circles). Based on the given psychophysical responses, the underlying logit link function predicts a 50 % SRT of about 3 dB (red cross and corresponding dashed lines). (Figure created using Jones, 2021.)

ing (Taylor & Creelman, 1967; Pentland, 1980), to maximum-likelihood procedures (Green, 1993; Leek, 2001). Various factors including the material of the target speech (phonetic balance, level of predictability, monosyllabic word lists vs. sentences), the noise type (reversed speech, speech-shaped noise, energetic/modulation/informational masking), or distractor features in relation to the target (e.g. pitch/gender differences, co-located/spatially separated) influence the individual SRT (see, e.g. Bronkhorst, 2000).

2.9.2 Spatial release from masking

A number of previous studies in different investigation groups reported improved speech perception and correspondingly reduced SRTs if a target talker is spatially separated from the interfering noise sources or distractors, compared to the scenario with co-located target and noise sources or distractors (see, e.g. Ching et al., 2011; Misurelli & Litovsky, 2015; Viveros Munoz, 2019). This benefit is commonly referred to as spatial release from masking (SRM) and described by Kock (1950) and Litovsky (2012). Figure 2.13 schematically depicts a symmetric scenario with a target talker T and two speech distractors D_1 and D_2. The two distractors are either spatially co-located with or symmetrically displaced from the target talker, which is located at 0 deg azimuth in the horizontal plane, by

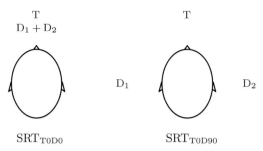

Figure 2.13: SiN test scenarios to measure the SRTs with varying distractor location. Left: Co-located arrangement of a target talker T and the two distractors D_1 and D_2 (SRT_{T0D0}), all located in the horizontal plane at 0 deg azimuth. Right: Symmetric distractor configuration with D_1 and D_2 angularly shifted by 90 deg azimuth to the left and right, while maintaining the original target talker location at 0 deg azimuth (SRT_{T0D90}).

some angle, for example ±90 deg, in the horizontal plane. SRM is then quantified by

$$SRM = SRT_{T0D0} - SRT_{T0D90}, \tag{2.18}$$

in dB, with SRT_{T0D0} and SRT_{T0D90} representing the SRTs measured in the scenarios with co-located and spatially separated target talker and distractors, respectively. Similar to the SRTs, the amount of SRM is influenced by factors including the target speech material, reverberation and HL (Marrone, Mason, & Kidd, 2008), as well as by the number, configuration, spectrum, type, distance and movement of maskers (for a review, see Viveros Munoz, 2019, Ch. 3).

3

Virtual acoustic environments for hearing aid research

The term "auralisation" describes the process of making numerical data, created by means of simulations, measurements or syntheses, audible via a sound file (Vorländer, 2020). With the aim of gradually bringing VAEs closer to a result that is indistinguishable from the intended real-world environment, research in the field of virtual acoustic reality has made considerable progress in recent decades. In this regard, physically based room acoustic simulations and dynamic real-time reproduction represent key components to increase the level of plausibility of virtual indoor scenes. This chapter starts with potential application areas and general requirements of VAEs. Thereafter, a novel concept on expanding such VAEs to address HA applications is introduced. The simulation and spatial audio reproduction methods are adapted accordingly, enabling proper integration of research HAs in the virtual scene. Finally, specific implementation variants of a combined binaural real-time auralisation system and a combined hybrid static auralisation system, both combining established spatial audio reproduction methods with HA-based reproduction, are presented.

3.1 Application areas

VAEs are suitable for a wide range of application areas including auditory research, HA algorithm development, HA fitting, clinical studies, auditory training, and multimodal research. Some examples are provided below.

Related publications:

Pausch et al., 2018b; Pausch and Fels, 2020, 2019b.

One of the most obvious advantages of VAEs is connected to the possibility of manipulating experimental conditions including their order, the number and poses of VSSs, and room acoustic parameters. In the corresponding real-world experiment, such experimental modifications would require elaborate physical changes of the related components (e.g. loudspeaker poses, absorber panels, etc.). In addition to the effort involved, the question remains whether any inaccuracies in the modification work, fluctuating ambient conditions and other external variables would introduce a latent bias to the experimental results and impair their internal validity. Control of extraneous factors is always a challenging task, for example in field studies (Braat-Eggen et al., 2017; Klatte et al., 2010a), and can also be improved through the application of VAEs.

Numerous studies have investigated the detrimental impact of noise, ranging from its disruptive potential in sensitive workplaces on cognitive performance and attention (Szalma & Hancock, 2011; Hellbrück & Liebl, 2008; Klatte et al., 2010b; Shafaghat et al., 2014; Marsh et al., 2018), over effects of aircraft noise (Hygge, Evans, & Bullinger, 2002), to noise-induced HL among elderly people (Nelson et al., 2005), which is sometimes aggravated by recruitment, hyperacusis or tinnitus (Axelsson & Sandh, 1985; Nelson & Chen, 2004). When such noise studies are to be conducted in VAEs, the simulation environment must allow to specify the types, locations, movements, as well as physical properties of the noise sources and their environment (Dreier & Vorländer, 2021). The possibility of controlling the emitted SPLs is crucial to prevent any short-term or long-term damage of study participants.

SiN perception plays an important role for language development, learning success, developing self esteem and general social competence (Yoshinaga-Itano et al., 1998; Theunissen et al., 2014). Related studies measured SRTs and SRM as performance indicators of speech perception in complex environments in the presence of maskers (Bronkhorst, 2000; Best, Mason, & Kidd, 2011; Westermann & Buchholz, 2015). Experimental variations included changing target and masker materials, and the number of distractors (Hawley, Litovsky, & Culling, 2004; Bronkhorst & Plomp, 1992), to address the effects of adverse room acoustics (Kidd et al., 2005; Culling, 2016) on sequential bilingual children (MacCutcheon et al., 2018) and children with HL (Marrone, Mason, & Kidd, 2008; Pausch et al., 2016c). Other investigations assessed the influence of radially or circularly moving maskers (Davis, Grantham, & Gifford, 2016; Viveros Munoz, 2019). The work by Cameron, Dillon, and Newall (2006) demonstrates the possibility of detecting central auditory processing disorders by applying SiN tests as part of clinical anamnesis strategies. Auditory training may subsequently serve as an intervention strategy to improve SiN performance (Cameron, Dillon, & Newall, 2006; Henshaw & Ferguson, 2013; Cameron, Glyde, & Dillon, 2012).

The prescription of HAs also represent an effective tool to overcome the negative effects associated with HL. However, evidence exists that people are not satisfied with the performance of their devices (Hougaard & Ruf, 2011) even after consulting an audiologist with the aim of finding an optimal HA fit, determined by optimally parametrised algorithms and settings. This indicates that typically used fitting routines with discrete loudspeakers reproducing simplistic speech in the presence of broadband noise maskers, sometimes conducted in a hearing booth, may not be sufficient to replicate complex real-life situations (see, e.g. Nikles & Tschopp, 1996). Objective evaluations of HA algorithms in VAEs help to understand reasons for this discrepancy by being able to increase the complexity of the laboratory condition, which allows to estimate and predict a real-world benefit (Walden et al., 2000; Cord et al., 2002; Compton-Conley et al., 2004). For indoor environments, room acoustics play an important role as additional reflections reduce the effectiveness of binaural cues (Plomp, 1976) and the performance of HA algorithms (Kates, 2001).

The application of VAEs is not limited to the auditory domain but may also be extended to interdisciplinary and multimodal research including the visual domain. Studies in the related fields include, for example, investigations on the influence of dynamic speaker directivity on the perception of speech (Ehret et al., 2020), and on movement and gaze behaviour in audiovisual environments (Hendrikse et al., 2019). It is also of particular interest, how a variation of the amount of visual information, conveyed via head-mounted displays, influences the localisation performance in humans (Ahrens et al., 2019). Other interdisciplinary research focuses on selective auditory attention and attention switching (Ruggles & Shinn-Cunningham, 2011; Oberem, Koch, & Fels, 2017), making extensive use of VAEs. Speech decoding in complex situations can be complementarily analysed on the basis of electroencephalogram data (Fuglsang, Dau, & Hjortkjær, 2017) or functional near-infrared spectroscopy (Bell et al., 2020), requiring strict synchronisation of temporal events during paradigm execution for event-related data analysis strategies.

3.2 Concept of a hearing aid auralisation system

To enable their application in the aforementioned research areas and particularly in HA-related studies, auralisation systems need to fulfil certain requirements to facilitate flexible and plausible simulation and reproduction of VAEs. These requirements are presented below and conceptualised by means of a hearing aid auralisation (HAA) system, presented in Figure 3.1.

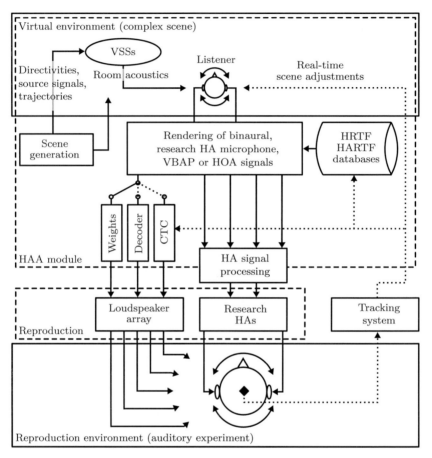

Figure 3.1: Concept of an HAA system. On top, a virtual environment for the generation of a complex indoor scene containing various elements is shown. The scene includes a number of static or moving VSSs that are linked to arbitrary source signals and characteristic directivities, to be perceived by a listener fitted with research HAs in simulated room acoustics. Following the rationale of combined perception via HAs and an external sound field, the respective HA microphone and ear signals are rendered accordingly. After processing the HA microphone signals, the receivers of the research HAs are used for playback. External sound field reproduction in the reproduction environment is accomplished by means of a loudspeaker array with appropriate signal conditioning. Real-world user movements are motion-tracked and fed back to the HAA module triggering real-time scene updates. (Figure taken from Pausch et al., 2018b, licensed under CC BY 4.0, and adapted.)

3.2.1 Simulation of complex acoustic scenes

The overall goal of a VAE is to recreate its real-world counterpart by conveying all acoustic cues in a plausible or even authentic manner. Authentic VAEs dictate highest standards regarding physical correctness and perception, since they have to be identical with the real-world equivalent, representing the external reference (Blauert, 1997). More relaxed requirements apply to plausible VAEs which need to satisfy the expectation of a listener, specifying a corresponding individual inner reference (Lindau & Weinzierl, 2012).

On top of Figure 3.1, a listener is immersed in a virtual scene and located at some current pose. Since this virtual scene may involve multiple VSSs, that are either static or moving along trajectories, it is classified as complex. Each VSS is linked to a specific source signal and characterised by its sound power level (SWL) and a distance-dependent SPL (Vorländer, 2020). Depending on the source signal, measured or simulated directivities further characterise the VSSs (e.g. Halkosaari, Vaalgamaa, & Karjalainen, 2005; Shabtai et al., 2017). Similarly, generic or individual directivities are assigned to the receiver, which can be measured from artificial heads (Algazi et al., 2001; Schmitz, 1995; Gardner & Martin, 1995; Brinkmann et al., 2017; Thiemann & van de Par, 2019), from individuals (Bomhardt, de la Fuente Klein, & Fels, 2016; Denk et al., 2018; Richter & Fels, 2019), individualised (Middlebrooks, 1999; Bomhardt, 2017; Pausch, Doma, & Fels, 2021), or numerically simulated (Katz, 2001; Fels, Buthmann, & Vorländer, 2004; Ziegelwanger, Majdak, & Kreuzer, 2015). The combination of movements including relative distance changes between VSSs and the receiver lead to time-varying physical effects like Doppler shifts which also need to be addressed (Strauss, 1998; Stienen, 2022). If the complex scene is situated in an indoor environment a corresponding three-dimensional model can be created using computer-aided design software for room acoustic simulations (Pelzer et al., 2014). By specifying the properties of the surface materials in this model and linking them with databases of acoustic parameters, such as absorption and scattering coefficients, the room acoustic simulation models are supported by physical principles, which increases their plausibility (Schröder, 2011; Kuttruff, 2016; Vorländer, 2020).

3.2.2 Signal rendering

Based on the physically based room model and the characteristics assigned to the VSSs and the receiver, different types of signals need to be rendered. As introduced in Section 2.8, individuals fitted with HAs perceive a spatialised sound source as a mixture of direction-dependent HA-based signals and via residual hearing. The directional receiver characteristics for each perceptual path can be stored as databases of rendering HARTFs and HRTFs in commonly used directional audio file formats, such as SOFA (Majdak et al., 2013) or OpenDAFF (2018). The directional transfer functions between each VSS and receiver are characterised by the corresponding IRs. To binaurally simulate the HA microphone signals, the IRs measured at the HA microphone positions including room acoustic filters, are convolved with an anechoic single-channel audio file associated with the VSS, resulting in a binaural signal, see Section 3.3.2. Depending on the spatial audio reproduction method used for external sound field reproduction, the rendering signals are accordingly generated, indicated by the switch in Figure 3.1, and further described in Sections 3.2.4 and 3.3.

3.2.3 Hearing aid signal processing and reproduction

In real life, the HA microphone signals are further processed via proprietary HA algorithms on the DSPs of the HAs. For economic and competitive reasons, it is obvious that manufacturers cannot disclose the specific implementation of HA algorithms. However, scientific studies and research applications require full access to parameters of HA algorithms. Even when using HAs from the same manufacturer, different HA models of the product range of a certain type, different fitting strategies and individual parameter settings of the involved algorithms may impact the results of perceptual experiments, additionally preventing unbiased inter-participant comparisons. For these reasons, the simulated binaural HA microphone signals are used as input signals to specifically designed HA software platforms (e.g. Curran & Galster, 2013; Herzke et al., 2017), that can remedy the situation by acting as an HA signal processing environment, referred to as master hearing aid (MHA). The number of input channels corresponds to the number of HA microphones to be further processed, typically lying in the range of two to six for a pair of HAs. It is necessary that such HA software platforms provide established fitting protocols and basic HA algorithms, such as beamforming, wide dynamic range compression or noise suppression, see Figure 2.7. Although these algorithms do not necessarily have to exhibit the same complexity and performance as algorithms in premium HAs, comprehensive and transparent parameter configuration possibilities of the MHA are crucial. The selection and configuration of the HA fitting should preferably be performed in cooperation with audiologists. Only by being able to set and control all components of the simulation chain is it possible to avoid any bias due to unknown behaviour of HA algorithms and across HAs from different manufacturers, thus allowing reproducible auditory research results. The output signals of the MHA are finally reproduced by the HA receivers in addition to the aforementioned external sound field reproduction. The HAs must feature the possibility to directly insert these reproduction signals, bypassing any HA algorithms, and forward them to the

HA receiver. To account for the signal processing delay in HAs (Stone et al., 2008), the HA-related signal path needs to feature the possibility to set Δt_{HA} in a way to achieve a desired relative delay between the HA receiver signals and the signals arriving at the blocked ear canal entrance, cf. Figure 2.8.

3.2.4 External sound field reproduction

If acoustic crosstalk cancellation (CTC) is applied for external sound field reproduction, classifying the system as entirely binaural, the calculation of the corresponding CTC filters relies on a playback HRTF database, which can be identical or different from the rendering HRTF database. Such CTC filters are used to filter and recreate the binaurally rendered source signal including room acoustic filters at the ear canal entrances. The loudspeaker array used for binaural playback consists of at least two properly arranged loudspeakers (Atal, Hill, & Schroeder, 1966; Bauer, 1961; Lentz, 2008; Parodi & Rubak, 2010) in the reproduction environment. Fundamentally different, the external sound field can also be recreated in hybrid system variants by the application of panning-based concepts, such as vector base amplitude panning (VBAP) or higher-order Ambisonics (HOA), requiring a spherical loudspeaker array that surrounds the listener. Before the simulated VBAP or HOA signals can be reproduced, a set of loudspeaker weights or a decoder matrix needs to be mapped onto the corresponding set of loudspeakers, respectively (Zotter & Frank, 2019). Note that in hybrid reproduction variants, the listener always applies individual receiver directivities to perceive the external sound field. The signal processing steps necessary for the rendering and reproduction of VAEs using these spatial audio methods are compactly introduced in Section 3.3.

3.2.5 Real-time rendering

Particularly in binaural auralisation systems, aiming to preserve localisation cues contained in HRTFs and HARTFs, see Sections 2.8.1 and 2.8.2, respectively, it is crucial to know the current real-world head pose of a listener to facilitate real-time scene updates. Motion tracking systems enable to estimate this pose and the relative directions and distances to the VSSs. At least parts of the simulation components of a scene must be updated during run time, aiming to maintain the VSS position and allow for a plausible perception of virtual scene elements. To avoid spatial discontinuities of VSSs in case of listener movements, the rendering and playback HRTFs and HARTFs must cover directions measured on a dense spatial grid. The requirement of low latency places high demands on the real-time engine, filter update rates, underlying simulation models, and the use of computational resources, particularly when involving a high number of VSSs. When it comes to the real-time simulation of virtual indoor scenarios, room acoustic simulations play an integral part to achieve a plausible listening experience. In order not to make the application of the auralisation system dependent on a server-based processing backbone, particularly when simulating long RTs requiring long filters, but rather allow its use on desktop computers with much more limited resources, efficient geometric room acoustic simulation models (Kuttruff, 2016; Vorländer, 2020) with perception-driven filter update rates need to be included and specified, respectively (Aspöck et al., 2014).

3.2.6 System latency

EEL is a crucial factor for real-time auralisation systems as it correlates with presence (Slater et al., 2009). The level of presence achieved determines the extent to which people respond to the sensory data generated by VAEs and interact accordingly with these environments just as they would do if these environments were real (Sanchez-Vives & Slater, 2005). In the scope of the current investigations, dynamic EEL is defined as duration between the time instance of a real-world user pose change and the time instance when the sound from the updated synthesised and reproduced acoustic scene reaches the ear drums of the listener. It is necessary that this time interval lies below just detectable thresholds of 60 to 75 ms on average (Brungart, Simpson, & Kordik, 2005; Yairi, Iwaya, & Suzuki, 2006), depending on the source signal and measurement method. Lindau (2009) reported no influence of the auralised VAE (anechoic, reverberant) or the stimulus type (noise, music, speech) on the detectable thresholds, reporting increased pooled threshold values of 107.63 ± 30.39 ms ($\mu \pm \sigma$).

3.2.7 Reproduction environment

Reproduction environments must fulfil high acoustic standards with limited reflection potential of surfaces to allow minimal RTs and excellent speech intelligibility. This is particularly important for loudspeaker-based binaural reproduction, which is likely to suffer from decreased performance subject to detrimental room reflections (Ward, 2001; Sæbø, 2001; Kohnen et al., 2016) . Sufficiently high shielding from external noise sources favours low ANLs and minimal distortion and distraction during critical acoustic measurements and perceptual experiments, respectively. Rigorous MPANLs when conducting audiometries are for example specified by ANSI/ASA S3.1 (1999) and ISO 8253-1 (2010). If the proposed HAA system shall be used in the clinical context the amount of hardware should be kept reasonable low to be installable in a hearing booth with restricted space, favouring the use of the entirely binaural implementation variant with a low loudspeaker count. The use of the hybrid implementation variants, relying on spherical loudspeaker arrays surrounding the listener, typically go along with increased hardware demands and require substantially more space. Stable ambient conditions serve well-being, which is particularly important in experiments involving children. Therefore, special care must be taken by providing a comfortable seating possibility, using neutral colours for the room surfaces and acoustic absorbers to avoid visual distraction, as well as discretely installing all involved technical components. Closely related is the importance of a constant room temperature and fresh air supply which, if not controlled properly, may induce fatigue, drop in concentration or general discomfort.

3.3 Spatial audio rendering and reproduction methods

An overview of spatial audio rendering and reproduction methods, used for the specific implementation variants of the presented HAA system, the objective and perceptual system evaluations, and the clinical experiment in this thesis, is given below. Since a comprehensive description is out of scope, the reader is referred to the literature, without claim to completeness, on:

- binaural headphone reproduction (Møller, 1992; Blauert, 1997; Wightman & Kistler, 2005a; Sanches Masiero, 2012);
- acoustic CTC (Bauck & Cooper, 1996; Lentz, 2008; Sanches Masiero, 2012);
- VBAP and variants (Pulkki, 1999, 2001; Zotter & Frank, 2019);
- HOA (Daniel, 2000; Zotter, 2009a; Zotter & Frank, 2019).

3.3.1 Monophonic reproduction via loudspeakers

Although not addressing VAEs, the most simple approach for spatial audio reproduction is playing back a single-channel loudspeaker input signal $X(k)$ via the l-th loudspeaker, $l = 1, \ldots, N$, of an N-channel loudspeaker array with radius R surrounding the listener, from direction (φ_l, θ_l), see Figure 3.2.

Assuming a frequency-independent acoustic centre, the loudspeaker main axis represents the direction perpendicular to the loudspeaker front running through this specified acoustic centre, referred to as on-axis direction. The loudspeaker is usually oriented in way that the on-axis direction intersects with the centre of the listener's interaural axis. Given that the on-axis frequency response $H_{\mathrm{LS}}(k)$ is known, spectral deviations can be equalised by means of multiplying the inverse on-axis transfer function $H_{\mathrm{LS}}^{-1}(k)$ with the loudspeaker output signal $Y'(k)$. This loudspeaker EQ filter can, for example, be implemented as FIR filter (Oppenheim, Schafer, & Buck, 1999; Pausch, Behler, & Fels, 2020), ideally resulting in a loudspeaker on-axis transfer function with flat passband magnitude spectrum. Depending on the pose relative to the loudspeaker, a listener without HAs always

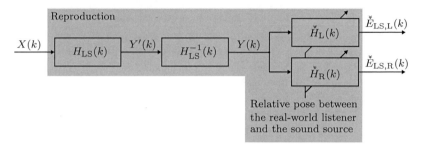

Figure 3.2: Monophonic signal reproduction via loudspeakers of a signal $X(k)$ filtered by the loudspeaker on-axis transfer function $H_{\mathrm{LS}}(k)$ and optionally equalised by including its inverse $H_{\mathrm{LS}}^{-1}(k)$. The loudspeaker signal $Y(k)$ is spatially filtered by the individual playback HRTFs $\breve{H}_{\{\mathrm{L,R}\}}(k)$, resulting in the signals $\breve{E}_{\mathrm{LS},\{\mathrm{L,R}\}}(k)$ at the ear drum.

utilises own individual HRTFs, represented by the playback HRTFs $\breve{H}_{\{\mathrm{L,R}\}}(k)$, to binaurally perceive the emitted loudspeaker signal as signals $\breve{E}_{\mathrm{LS},\{\mathrm{L,R}\}}(k)$ at the ear drum, which are further influenced by ear canal resonances. Note that $\breve{H}_{\{\mathrm{L,R}\}}(k)$ includes reflections from neighbouring array loudspeakers, the mounting construction, and the reproduction environment if the array is not installed in an anechoic chamber.

Discrete spatialised loudspeakers emulate RSSs and are commonly used as experimental reference condition in localisation experiments, as they allow highest localisation accuracy. As an example, Bronkhorst (1995) compared the localisation performance in this reference condition against the one measured when presenting VSSs. Given an array of loudspeakers, it is also possible to simulate reflections represented by individual loudspeakers using delayed and gain-adjusted loudspeaker signals (Seeber, Kerber, & Hafter, 2010). Seeber, Baumann, and Fastl (2004) used this approach to measure the localisation performance in the horizontal plane in listeners fitted with bimodal HAs or bilateral cochlear implants.

3.3.2 Binaural reproduction via headphones and hearing aids

An intuitive approach is reproducing a binaural signal via a set of headphones, see Figure 3.3a. The rendering stage involves binaural synthesis and requires to multiply the input signal $X(k)$ with a set of rendering HRTFs $H_{\{\mathrm{L,R}\}}(k)$, which depend on the pose of the virtual listener relative to the VSS to be reproduced, resulting in the left and right binaural signals $B_{\{\mathrm{L,R}\}}(k)$. This relative pose also determines the resulting playback level, for example, by utilising the inverse distance law for sound pressure decay in free-field conditions (Vorländer, 2020). For

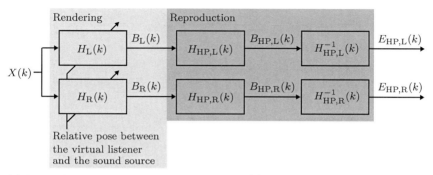

(a) Binaural signal rendering of a source signal $X(k)$ using a set of HRTFs $H_{\{L,R\}}(k)$ to obtain binaural signals $B_{\{L,R\}}(k)$. Before playback via headphones $H_{HP,\{L,R\}}(k)$, the binaural signals $B_{HP,\{L,R\}}(k)$ are individually corrected using a headphone EQ $H^{-1}_{\{L,R\}}(k)$, resulting in the perceived ear signals $E_{HP,\{L,R\}}(k)$.

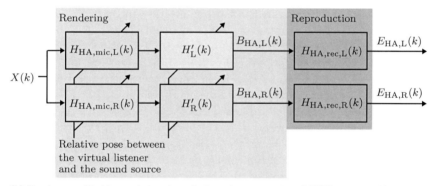

(b) Device-specific binaural signal rendering of a source signal $X(k)$, captured by a pair of HA microphones $H_{HA,mic,\{L,R\}}(k)$, e.g. the front microphones, and a set of HARTFs $H'_{\{L,R\}}(k)$. The binaural signals $B_{HA,\{L,R\}}(k)$ are reproduced via the HA receivers $H_{HA,rec,\{L,R\}}(k)$, resulting in the perceived ear signals $E_{HA,\{L,R\}}(k)$.

Figure 3.3: Binaural signal rendering and reproduction via **(a)** headphones and **(b)** research HAs.

increased perceptual plausibility (Oberem, Masiero, & Fels, 2016), the headphone transducer transfer functions $H_{HP,\{L,R\}}(k)$ need to be compensated for by individual measurements. For this purpose, a robust equalisation routine, proposed by Masiero and Fels (2011), was conducted prior to all headphone-based experiments in this thesis. The equalisation routine requires measuring the acoustic transfer paths between the left and right headphone transducers and the corresponding in-ear microphones (KE 3, Sennheiser, Wedemark, Germany), the influence of which must be removed by spectral division using their free-field response, placed at the blocked ear canal entrances. The measurements should ideally be repeated at least eight times, each time after repositioning the headphones, to obtain the upper variance limit of the set of measured headphone transfer functions, which is subsequently applied as inverse minimum-phase filter on the respective binaural signals $B_{\mathrm{HP},\{\mathrm{L,R}\}}(k)$, resulting in the binaural ear signals $E_{\mathrm{HP},\{\mathrm{L,R}\}}(k)$ at the ear canal entrance. Particular care must be taken not to change the overall loudness of the binaural signals but to only equalise their spectra. Example headphone transfer functions, measured from a set of circum-aural headphones fitted to one individual with variations, and the corresponding headphone EQ are shown in Pausch (2022, Fig. 6).

A schematic depiction of the binaural rendering and reproduction scenario using research HAs is shown in Figure 3.3b. In contrast to headphone listening, where the influence of the listening device must be compensated for, the HA microphone and receiver characteristics are an integral part during rendering and reproduction. Depending on the pose of the virtual listener relative to the VSS, the HA microphones $H_{\mathrm{HA,mic},\{\mathrm{L,R}\}}(k)$, i.e. either the front or rear pair, or all microphones as necessary for optionally involved HA algorithms, and the corresponding HARTFs, all with direction-dependent characteristics, see Pausch (2022, Sec. 3.4) and 2.8.2, are multiplied with the input spectrum $X(k)$. The resulting binaural signals $B_{\mathrm{HA},\{\mathrm{L,R}\}}(k)$ are subsequently reproduced via the HA receivers, including their specific frequency characteristics $H_{\mathrm{HA,rec},\{\mathrm{L,R}\}}(k)$ depending on the ear piece, see Pausch (2022, Sec. 3.4). The ear signals $E_{\mathrm{HA},\{\mathrm{L,R}\}}(k)$ are further influenced by resonances in the remaining part of the ear canal.

3.3.3 Binaural reproduction via loudspeakers

If binaural signals are supposed to be reproduced via loudspeakers the listener will perceive these signals as a mixture at the left and right ear, resulting in distorted or inaccessible binaural and monaural cues and a disrupted spatial impression, commonly referred to as acoustic crosstalk. This phenomenon occurs because the binaural loudspeaker signals do not only reach the dedicated ears, as it is the case during binaural reproduction via headphones with nearly perfect CS, neglecting possible inter-cranial transaural transmission (Stenfelt, 2012), but are partly perceived on the unwanted ear. An optimised filter network of CTC filters, applied to the binaural input signal, aims to suppress these acoustic crosstalk paths while preserving the binaural signal at the desired ear to achieve a sufficiently high CS at a specific real-world listener pose in the reproduction environment.

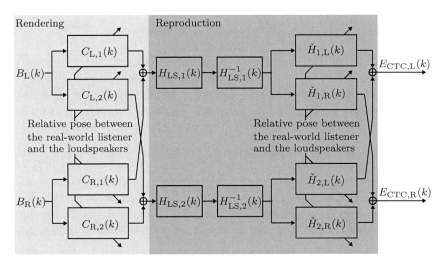

Figure 3.4: Rendering of a binaural signal for reproduction via loudspeakers applying CTC filters.

Figure 3.4 schematically presents the rendering and reproduction components of a CTC system consisting of $N' = 2$ loudspeakers. The framework developed by Sanches Masiero (2012) will be re-introduced below. The indices of the CTC filters and playback HRTFs largely follow the same nomenclature, defining the acoustic paths between the respective loudspeakers ($\{1,2\}$) and the ear sides ($\{L,R\}$). The binaural input signals $B_{\{L,R\}}(k)$ is rendered analogously as done for headphone-based reproduction, see Section 3.3.2. Based on the real-world pose of the listener relative to the two loudspeakers, a set of playback HRTFs $\check{H}_{\{1,2\},\{L,R\}}(k)$, i.e. the acoustic binaural transfer functions between each loudspeaker to the left and right ears of the listener with current real-world pose in the reproduction environment, is selected. Note that $\check{H}_{\{1,2\},\{L,R\}}(k)$ represent time-windowed playback transfer functions, ideally not containing reflections from the experimental setup or the reproduction environment. Often, $\check{H}_{\{1,2\},\{L,R\}}(k)$ are selected from the database of spatial rendering transfer functions $H_{\{1,2\},\{L,R\}}(k)$. Assuming that all loudspeaker transfer functions are perfectly equalised, meaning that $H_{\mathrm{LS},\{L,R\}}(k) \cdot H_{\mathrm{LS},\{L,R\}}^{-1}(k) \equiv 1$, the ear signals at the ear canal entrances after filtering the binaural input signal using a set of CTC filters are defined as

$$\begin{bmatrix} E_{\mathrm{CTC,L}}(k) \\ E_{\mathrm{CTC,R}}(k) \end{bmatrix} = \begin{bmatrix} \check{H}_{1,\mathrm{L}}(k) & \check{H}_{2,\mathrm{L}}(k) \\ \check{H}_{1,\mathrm{R}}(k) & \check{H}_{2,\mathrm{R}}(k) \end{bmatrix} \cdot \begin{bmatrix} C_{\mathrm{L},1}(k) & C_{\mathrm{L},2}(k) \\ C_{\mathrm{R},1}(k) & C_{\mathrm{R},2}(k) \end{bmatrix} \cdot \begin{bmatrix} B_{\mathrm{L}}(k) \\ B_{\mathrm{R}}(k), \end{bmatrix} \quad (3.1)$$

or, equivalently, as

$$\mathbf{e}_{\mathrm{CTC}} = \check{\mathbf{H}}\mathbf{C}\mathbf{b}. \tag{3.2}$$

In an ideal CTC system, the CTC filter matrix $\mathbf{C}$ is able to cancel out the acoustic crosstalk paths entirely, so that the binaural input signals $\mathbf{b}$ are unaffected and perfectly reproduced at the listener's ears, which is true for $\check{\mathbf{H}}\mathbf{C} = \mathbf{I}$, with $\mathbf{I}$ being the identity matrix. This requires to find a CTC matrix (Sanches Masiero, 2012; Oppenheim, Schafer, & Buck, 1999)

$$\mathbf{C} = \check{\mathbf{H}}^{-1} e^{-j2\pi k n_0/K}, \tag{3.3}$$

that not only aims at properly inverting the playback HRTF matrix but is also causal. Causality is achieved by introducing a discrete time delay n_0, addressing the acoustic runtime between the real-world listener position and the loudspeakers to avoid echoes or artefacts due the cyclic property of the DFT, see Section 2.2.3 and Sanches Masiero (2012).

Acoustic CTC was originally introduced as a transaural stereo system (Bauer, 1961; Atal, Hill, & Schroeder, 1966; Bauck & Cooper, 1996). Lentz (2008) demonstrated that the CTC filters in such systems will become instable as soon as the real-world viewing angle of the listener approaches or lies within a critical angular range of the respective loudspeaker. He therefore proposed dynamic selection of loudspeaker pairs according to the listener real-world pose and defined spatial regions to crossfade between the pairs, requiring at least three or four loudspeakers. However, binaural reproduction is not limited to one or more loudspeakers pairs but can be performed using an array of $N' > 2$ simultaneously active loudspeakers, a technique referred to as N'-CTC. Consequently, this approach requires expanding the playback HRTF matrix to (Sanches Masiero, 2012)

$$\check{\mathbf{H}} = \begin{bmatrix} \check{H}_{1,\mathrm{L}}(k) & \check{H}_{2,\mathrm{L}}(k) & \dots & \check{H}_{N',\mathrm{L}}(k) \\ \check{H}_{1,\mathrm{R}}(k) & \check{H}_{2,\mathrm{R}}(k) & \dots & \check{H}_{N',\mathrm{R}}(k) \end{bmatrix}. \tag{3.4}$$

Substituting Equation (3.4) into Equation (3.3) and given one listener represents an underdetermined constrained optimisation problem (Nelson & Elliott, 1995; Sanches Masiero, 2012). The combination of additional measurement noise and a poorly conditioned matrix $\mathbf{H}$ may lead to problems regarding numerical stability and invertibility, generally hindering to find an acceptable optimal and stable solution. To address this issue and to avoid excessively large numerical values, resulting in unpleasant narrowband spectral CTC filter gains, an inverse system matrix, optimal in the least-squares sense with Tikhonov regularization (Farina, 2007), can be used, yielding (Sanches Masiero, 2012)

$$\mathbf{C} = \check{\mathbf{H}}^{-1} \approx \check{\mathbf{H}}^{\mathsf{H}}(\check{\mathbf{H}}\check{\mathbf{H}}^{\mathsf{H}} + \beta\mathbf{I})^{-1} \cdot e^{-j2\pi k n_0/K}, \tag{3.5}$$

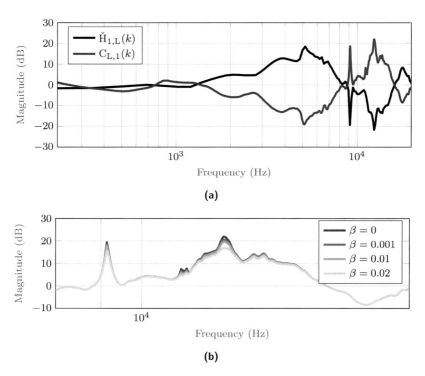

Figure 3.5: (a) Example left-ear playback HRTF from a loudspeaker located at $(\varphi_1, \theta_1) = (45°, 70°)$ with corresponding CTC filter without regularisation $(\beta = 0)$. **(b)** Influence of the regularisation parameter β on the magnitude spectrum of the example CTC filter.

with $(\cdot)^{\mathsf{H}}$ symbolising the Hermitian transpose, and β being a constant regularization parameter (e.g. $\beta = .01$). This CTC matrix, representing a regularised right-inverse, is subsequently applied in Equation (3.2). In practical system implementations, the distance of the used loudspeakers shall be large in relation to the loudspeaker dimensions (Guang et al., 2016). Parodi and Rubak (2010) additionally suggested to install the loudspeakers at elevated positions to favour less pronounced notches and azimuthal variations in playback HRTFs, and thus a better conditioned system matrix $\check{\mathbf{H}}$ prior to inversion.

Figure 3.5a shows the magnitude spectra of the left-ear playback HRTF of the left loudspeaker, $H_{1,\mathrm{L}}(k)$, in a 2-channel loudspeaker setup at $(\varphi_1, \theta_1) = (45°, 70°)$ and $(\varphi_2, \theta_2) = (315°, 70°)$. The resulting CTC filter $C_{\mathrm{L},1}(k)$ aims at compensating this ipsilateral transfer path – in combination with the not shown CTC filter $C_{\mathrm{L},2}(k)$ – to achieve a flat magnitude spectrum, thus exhibiting the approximate inverse shape of the playback HRTF. Since no regularisation had been applied the narrowband dips in the playback HRTF magnitude result in sharp peaks in the CTC filter. As shown in Figure 3.5b, these peaks are gradually decreased by increasing the regularisation parameter β.

Acoustic CTC systems fall in the category of authentic reproduction systems as they allow physically based broadband binaural signal reproduction. For experiments involving participants with HL, placing the HAs under headphones would potentially lead to acoustic feedback issues and uncontrolled behaviour of HA algorithms, thus motivating the use of such "virtual headphones". The virtualisation not only results in a reduced restriction of movement due to the absence of the headphones but also favours an increased level of immersion. Ideally, both rendering and playback transfer functions are matched to individual HRTFs, as otherwise a substantial decrease in localisation performance is likely to be observed (Majdak, Masiero, & Fels, 2013).

3.3.4 Vector base amplitude panning

In contrast to the introduced physically correct binaural reproduction techniques, panning-based approaches are classified as being plausible (VBAP) or physically correct in a restricted frequency range (HOA). In classical stereo panning methods, a VSS is reproduced as a phantom source within the span angle between two loudspeakers, utilising differences in signal levels and TOAs between the loudspeaker signals. Wendt (1963) investigated the influence of variations in these factors on the perception of VSSs in the horizontal plane using a variety of transient and band-limited sounds, resulting in characteristic panning curves.

Pulkki (1997) proposed an extension of two-dimensional panning to include possible directions of incidence in three-dimensional space by selecting a single loudspeaker, pairs, or triplets from the set of N loudspeakers, which are, for example, part of a spherical array surrounding the listener. The centre of the array is typically coinciding with the origin of the global coordinate system, see Section 2.1. To pan the VSS to a certain direction $\theta_p = [x_p \quad y_p \quad z_p]^\mathsf{T}$, optimised amplitude weights $\mathbf{g}$ are applied on the respective signals of the involved loudspeakers. In the horizontal plane, only single loudspeakers or pairs will be selected, i.e. the number of active loudspeakers N′ equals 1 or 2, whereas in three-dimensional space between one and three loudspeakers may be involved, i.e. N′ equals 1, 2 or 3. Since only a subset N′ of loudspeakers is active, such approaches are classified as local panning approaches (Spors et al., 2013).

The underlying rationale aims to reproduce a VSS as close as possible from a desired panning direction θ_p. This panning direction is spatially enclosed by the involved loudspeakers and defined as a weighted sum of loudspeaker direction vectors, represented by unit-length vectors $\boldsymbol{\theta}_l = \begin{bmatrix} x_l & y_l & z_l \end{bmatrix}^{\mathsf{T}}$ pointing at the respective loudspeaker, yielding (Pulkki, 1997; Zotter & Frank, 2019)

$$\boldsymbol{\theta}_{\mathrm{p}} = \sum_{l=1}^{\mathrm{N}'} \tilde{g}_l \boldsymbol{\theta}_l. \tag{3.6}$$

To avoid loudness variations owing to changes in the number of involved loudspeakers, the weights $\tilde{g}_l$ are normalised, yielding

$$g_l = \frac{\tilde{g}_l}{\sqrt{\sum_{l=1}^{\mathrm{N}'} \tilde{g}_l^2}}. \tag{3.7}$$

In matrix notation, these sets of linear combinations including, for example, three loudspeaker direction vectors can be compactly written as (Pulkki, 1999; Zotter & Frank, 2019)

$$\boldsymbol{\theta}_{\mathrm{p}} = \begin{bmatrix} \boldsymbol{\theta}_1 & \boldsymbol{\theta}_2 & \boldsymbol{\theta}_3 \end{bmatrix} \begin{bmatrix} \tilde{g}_1 \\ \tilde{g}_2 \\ \tilde{g}_3 \end{bmatrix} = \boldsymbol{\Theta} \cdot \tilde{\mathbf{g}} \tag{3.8}$$

$$\Rightarrow \quad \tilde{\mathbf{g}} = \boldsymbol{\Theta}^{-1} \boldsymbol{\theta}_{\mathrm{p}}, \text{ and } \mathbf{g} = \frac{\tilde{\mathbf{g}}}{\|\tilde{\mathbf{g}}\|_2}, \tag{3.9}$$

with $\|\cdot\|_2$ representing the L2 vector norm. As long as the loudspeaker direction vectors span a three-dimensional space, an inverse $\boldsymbol{\Theta}^{-1}$ exists.

Vector base intensity panning

Closely related but using a set of squared weights to improve the angular mapping and its perception, the panning direction $\boldsymbol{\theta}_{\mathrm{p}}$ using vector base intensity panning (VBIP) is defined as (Zotter & Frank, 2019)

$$\boldsymbol{\theta}_{\mathrm{p}} = \begin{bmatrix} \boldsymbol{\theta}_1 & \boldsymbol{\theta}_2 & \boldsymbol{\theta}_3 \end{bmatrix} \begin{bmatrix} \tilde{g}_1^2 \\ \tilde{g}_2^2 \\ \tilde{g}_3^2 \end{bmatrix} = \boldsymbol{\Theta} \cdot \tilde{\mathbf{g}}_{\mathrm{sq}} \tag{3.10}$$

$$\Rightarrow \quad \tilde{\mathbf{g}}_{\mathrm{sq}} = \boldsymbol{\Theta}^{-1} \boldsymbol{\theta}_{\mathrm{p}}, \quad \tilde{\mathbf{g}} = \begin{bmatrix} \sqrt{\tilde{g}_{\mathrm{sq},1}^2} \\ \sqrt{\tilde{g}_{\mathrm{sq},2}^2} \\ \sqrt{\tilde{g}_{\mathrm{sq},3}^2} \end{bmatrix}, \text{ and } \mathbf{g} = \frac{\tilde{\mathbf{g}}}{\|\tilde{\mathbf{g}}\|_2}. \tag{3.11}$$

Energy metrics and vectors

Gerzon (1992) introduced the sum of squared loudspeaker gains, $\sum_{l=1}^{\mathrm{N}'} g_l^2$, as a perceptual correlate for direction-dependent loudness. The directional mapping

performance of panning-based reproduction methods can be further quantified
on the basis of the velocity vector (Makita, 1962)

$$r_V = \frac{\sum_{l=1}^{N'} g_l \boldsymbol{\theta}_l}{\sum_{l=1}^{N'} g_l}, \tag{3.12}$$

the direction of which supports localisation below frequencies of about 700 Hz.
For higher frequencies, the panning-dependent spatial energy accumulation, represented by the energy vector (Gerzon, 1992)

$$r_E = \frac{\sum_{l=1}^{N'} g_l^2 \boldsymbol{\theta}_l}{\sum_{l=1}^{N'} g_l^2} \tag{3.13}$$

and its magnitude, can be binaurally evaluated, helping to localise a VSS and
correlating with its apparent source width (Zotter & Frank, 2019; Gerzon, 1992).

Loudspeaker selection

The determination of active triplets relies on complex hull algorithms (Preparata
& Shamos, 1985) like, for example, Delaunay triangulation (Delaunay et al., 1934).
Depending on $\boldsymbol{\theta}_p$, a search algorithm subsequently selects a suitable loudspeaker
triplet with all amplitude weights being positive. Loudspeaker span angles greater
than 90 deg will likely entail directional instability of the reproduced VSS, or even
lead to sets of loudspeaker gains containing only zeros, and should therefore be
avoided (Pulkki, 2001; Zotter & Frank, 2019). Figure 3.6a shows the Delaunay
triangulation for a 16-channel loudspeaker setup, cf. Pausch (2022, Sec. 3.6 and
Fig. 12), and the active loudspeaker triplet with the corresponding loudspeaker
gains to reproduce a VSS from an example direction, indicated by the red cross.

Multiple-direction amplitude panning

The varying number of loudspeakers has the adverse effects of direction-dependent
changes in the apparent VSS width and colouration artefacts that are perceivable
particularly in case of moving VSSs (Pulkki, Karjalainen, & Välimäki, 1999;
Frank, 2013). To remedy this issue, Pulkki (1999) proposed multiple-direction
amplitude panning (MDAP), a technique which introduces auxiliary sources
to reduce the fluctuation in the number of active loudspeakers at the cost of
an increased apparent source width. These auxiliary sources are arranged in
equiangular steps on a concentric ring around $\boldsymbol{\theta}_p$, each of which needs to be
auralised in addition to $\boldsymbol{\theta}_p$. The by-loudspeaker sets of weights per source are
summed up and normalised according to Equations (3.9) or (3.11). In Figure 3.6b,

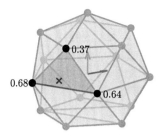

(a) Active loudspeaker triplet and corresponding common face represented by black dots and dark grey shading, respectively, including the loudspeaker weights.

(b) Introduction of 10 auxiliary sources at a spreading angle of 20 deg, requiring an additional loudspeaker in the lowest ring.

Figure 3.6: Examples of **(a)** VBAP and **(b)** MDAP configurations based on a triangu-lation of a regular loudspeaker layout (grey dots) to reproduce a VSS from direction $(\varphi, \vartheta) = (215°, 10°)$, indicated by the red cross. The nominal view-ing direction is visualised by view and up vectors in the centre of the array. (Gains and auxiliary source directions were calculated using Politis, 2021b).

the same VSS example direction as selected for VBAP is simulated using one ring of auxiliary sources.

Regardless of the used vector base panning technique, a single-channel audio file can be encoded as VSS using appropriate VBAP, VBIP or MDAP loudspeaker weights and reproduced as long as $\boldsymbol{\theta}_\mathrm{p}$ lies within the angular range covered by the regular array loudspeaker layout surrounding the listener and as long as the loudspeaker span angles do not exceed the critical angle of 90 deg. If the latter condition is violated the insertion of imaginary loudspeakers allows to address unstable angular ranges (Zotter & Frank, 2012, 2019).

The listener will binaurally perceive the VSS using individual HRTFs, sim-ilarly as when played back via a single loudspeaker, cf. Section 3.3.1, but as a summed contribution of unwindowed playback HRTFs $\breve{H}_{\{\mathrm{VBAP,MDAP}\},\{\mathrm{L,R}\}}(k)$ between each active loudspeaker and the respective ear, referred to as ear signals $\breve{E}_{\{\mathrm{VBAP,MDAP}\},\{\mathrm{L,R}\}}(k)$. A distortion of monaural and binaural cues can be ex-pected and will introduce localisation errors of simulated VSSs, for example, in the saggital plane (Baumgartner & Majdak, 2015). However, this approach can be efficiently used to recreate VAEs in a plausible way (Savioja et al., 1999), with the possibility to include room acoustic simulations (Pelzer, Masiero, & Vorlän-der, 2014). To reduce the influence of colouration (Frank, 2013), a simulation via discrete loudspeakers applying nearest neighbour panning would be a conceivable alternative.

3.3.5 Higher-order Ambisonics

For spatial audio reproduction using HOA, a set of N loudspeakers is arranged on an optimised regular layout (Zotter, 2009b) to sample a sphere with constant radius R, concentric to the nominal listener position. The sound field at the nominal listener position is synthesised by a number of weighted and superimposed orthonormal basis functions (Fellgett, 1975; Gerzon, 1975, 1985), facilitating the use of a continuous virtual panning function to angularly map a previously encoded source signal and reproduce it as a plane wave from a desired panning direction θ_p (Zotter & Frank, 2019).

Spherical harmonics

Expressing the Helmholtz equation in spherical coordinates and separating its variables by means of a product ansatz results in three different ordinary differential equations, with general solutions for the radial, and the two angular components (Williams, 1999; Zotter, 2009a). The angular solutions in θ and φ can be compactly summarised by a set of real-valued orthonormal basis functions. These spherical harmonics (SHs) of order n and degree m are defined as (Meyer & Elko, 2016)

$$Y_n^m(\boldsymbol{\theta}) = \sqrt{\frac{(2n+1)}{4\pi} \frac{(n-|m|)!}{(n+|m|)!}} \cdot P_{n,|m|}(\cos(\theta)) \cdot$$

$$\begin{cases} \sqrt{2}\sin(|m|\varphi), & \text{for } m < 0, \\ 1, & \text{for } m = 0, \\ \sqrt{2}\cos(|m|\varphi), & \text{for } m > 0, \end{cases} \tag{3.14}$$

normalised as per N3D normalisation, with $(\cdot)!$ denoting the factorial, and $P_{n,|m|}$ representing the associated Legendre functions of the first kind, without Condon-Shortley phase (Williams, 1999; Abramovitz & Stegun, 1964).

Encoding and decoding

To obtain the HOA signals of an input signal $x(n)$ that should be recreated as a plane wave from the desired panning direction θ_p, the panning coefficients correspond to the set of $(\mathcal{N}+1)^2$ SHs evaluated for this panning direction, representing the encoder (Zotter & Frank, 2019):

$$\boldsymbol{\psi}_{\mathcal{N}} = \boldsymbol{y}_{\mathcal{N}}(\boldsymbol{\theta}_p)x, \tag{3.15}$$

with $\boldsymbol{y}_{\mathcal{N}}(\boldsymbol{\theta}_p) = [Y_0^0(\boldsymbol{\theta}_p), Y_1^{-1}(\boldsymbol{\theta}_p), \ldots, Y_{\mathcal{N}}^{\mathcal{N}}(\boldsymbol{\theta}_p)]^\mathsf{T}$ to be multiplied with each sample of x. The simplest decoder is defined by evaluating the SHs at the loudspeaker positions (Malham & Myatt, 1995; Zotter & Frank, 2012)

$$D = \sqrt{\frac{4\pi}{N}} \left[y_{\mathcal{N}}(\theta_1), \ldots, y_{\mathcal{N}}(\theta_N) \right]^{\mathsf{T}} = \sqrt{\frac{4\pi}{N}} \, Y_{\mathcal{N}}^{\mathsf{T}}, \tag{3.16}$$

with normalisation constant $\sqrt{4\pi/N}$, called sampling ambisonic decoder (SAD). The loudspeaker signals are subsequently calculated by

$$y_{\mathrm{HOA}} = D \operatorname{diag}\{a_{\mathcal{N}}\} \psi_{\mathcal{N}}, \tag{3.17}$$

including the gains

$$g = D \operatorname{diag}\{a_{\mathcal{N}}\} y_{\mathcal{N}}(\theta_{\mathrm{p}}). \tag{3.18}$$

In contrast to VBAP and variants, see Section 3.3.4, all N loudspeakers are involved, i.e. $N' = N$, thus classifying HOA as a global panning method (Spors et al., 2013). Although encoding and decoding are generally independent from each other, the SH vector in Equation (3.16) is often truncated at $\mathcal{N} = \lfloor \sqrt{N} \rfloor - 1 \ll \infty$ (Poletti, 2005). Similar as observed for truncated Fourier series expansions (Oppenheim, Schafer, & Buck, 1999), this necessary practical order truncation leads to unwanted main and side lobes in the resulting panning function. However, effective suppression of side lobes at the cost of an increased main lobe width can be achieved by suitable order-dependent weights, applied as square diagonal matrix $\operatorname{diag}\{a_{\mathcal{N}}\}$ in Equation (3.18). For the experiments in this thesis involving HOA, the approximated version of the max r_{E} weighting (Daniel, 2000; Zotter & Frank, 2012), that is

$$a_n = P_n\left(\cos\left(\frac{137.9°}{\mathcal{N} + 1.51} \right) \right) \tag{3.19}$$

is utilised, the effect of which is plotted for example maximum orders $\mathcal{N}$ in Zotter and Frank (2019, Fig. 4.13, p. 70). Apart from a decreased spatial resolution during reproduction, order truncation has the effect that the sound field synthesis is limited by an upper frequency limit, above which spatial aliasing occurs (Zotter, 2009b). A limited order additionally reduces the size of the sweet sphere, i.e. the valid sound field synthesis area represented by a concentric sphere around the nominal listener position (Poletti, 2005; Frank & Zotter, 2017).

Compared to the SAD, see Equation (3.16), the improved all-round ambisonic decoder (AllRAD+) represents a more advanced decoder strategy. Irregular sampling layouts, i.e. loudspeaker arrangements that do not cover the whole sphere or violate strict sampling schemes (Zotter, 2009b) may result in panning-variant loudness, particularly in regions with poor loudspeaker coverage. This applies, for example, to loudspeaker arrangements on a spherical cap, not covering angles close to the south pole, cf. setups in Pausch (2022, Sec. 3.6 and 3.7). Initiated by the work of Batke and Keiler (2010), Zotter and Frank (2012) proposed to decode the HOA signals in the first step for a set of virtual loudspeakers, based on a spherical t-design layout (Hardin & Sloane, 1996). For $t > 2\mathcal{N} + 1$, the SAD is optimal. In the second step, these signals are mapped on the N real loudspeakers via the corresponding matrix of VBAP weights, defining the decoder $\boldsymbol{D}_{\text{AllRAD}}$. To further improve constant directional loudness perception, an additional weighting of the two decoding stages has been proposed by Zotter, Frank, and Pomberger (2013), resulting in $\boldsymbol{D}_{\text{AllRAD+}} = 1/\sqrt{2}\,\boldsymbol{D}_{\text{AllRAD}} + 1/\sqrt{8}\,\boldsymbol{D}_{\text{SAD}}$. Since the VBAP gains become zero for panning directions lying in angular areas with enclosing loudspeaker angles larger than 90 deg, the insertion of imaginary loudspeakers allows to preserve even these directions at the cost of an improved localisation error and thus extend the angular range of VBAP (Zotter & Frank, 2012).

Analogous to VBAP and variants, the listener will binaurally perceive the VSS via individual HRTFs as the sum of contributions of unwindowed individual playback HRTFs, $\breve{H}_{\{\text{HOA}\},\{\text{L,R}\}}(k)$, between all N loudspeakers and the respective ear, resulting in the ear signals $\breve{E}_{\{\text{HOA}\},\{\text{L,R}\}}(k)$. Apart from the high panning flexibility and mathematically elegant formulation of the virtual panning function, a few shortcomings of HOA have been reported in related publications. For example, perceptual artefacts can be traced back to phase distortions (Daniel, 2000) or comb-filter effects (Frank, Zotter, & Sontacchi, 2008), whose impact also depend on the reproduction environment (Santala et al., 2009). Oreinos and Buchholz (2015) and Grimm, Kollmeier, and Hohmann (2016) performed specific evaluations of HOA tailored to HA research.

3.4 Specific implementation

Two specific implementation variants with binaural and panning-based external sound field simulation for real-time and static reproduction, respectively, as used for the experiments in this thesis, are presented below. The hardware requirements of the first setup was kept as low as possible to be installed in a hearing booth, thus allowing on-site clinical experiments. The second variant is more hardware-intense and requires a substantially larger, and ideally acoustically optimised, reproduction environment, rather limiting its use to specialised research facilities.

3.4.1 Combined binaural real-time auralisation

Aiming to fulfil the requirements introduced in Section 3.2, Figure 3.7 presents the major signal processing steps of the implemented binaural real-time aural-isation system with an interface to research HAs. The HAA module is shown

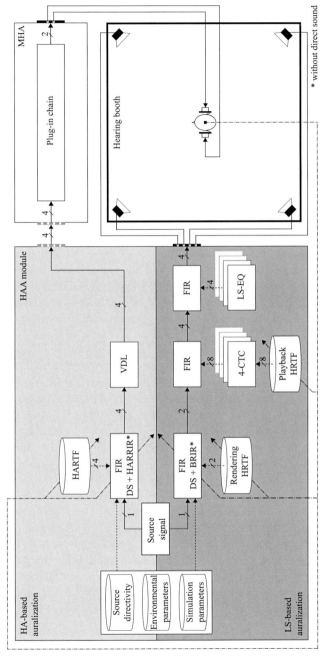

Figure 3.7: Specific implementation of the binaural real-time auralisation system extended to research HAs, presenting the signal flow in the HAA module to render the HA-related and loudspeaker-based signals, the integration of the MHA, and the combined reproduction in a hearing booth with real-time motion tracking feedback signals (dashed-dotted line). Signal filtering is implemented throughout with FIR filters. Parameter signals to update the FIR filter coefficients and spatial transfer functions are indicated as arrows with dashed lines. Software and hardware connections are shown as grey and black inlet/outlet boxes, respectively. (Figure taken from Pausch et al., 2018b, licensed under CC BY 4.0.)

on the left side with rendering components for the HA-based and loudspeaker-based signals. Referring to Figure 2.3, the different parts of the IR have to be addressed during simulation. The DS of the binaural signal in the corresponding auralisation path is simulated based on rendering HRTF and HARTF datasets, respectively. These spatial transfer functions were measured either from an artificial head (generic data) or from individuals (individual data), each time equipped with a pair of research HA, using the setups presented in Pausch (2022, Sec. 3.8 and 3.9), and stored in OpenDAFF (2018) format for efficient access. Early reflections and late reverberation are consolidated by hearing aid-related room impulse responses (HARRIRs) and binaural room impulse responses (BRIRs), both simulated using geometric room acoustic models as described in the upcoming section, to be merged with the respective HRTF and HARTF to an FIR filter. For increased plausibility and physical substantiation, databases containing individual source directivities are accessed. Environmental parameters such as temperature and humidity, and parameters related to the simulation can be set additionally. Any movements of VSSs are defined via pre-defined source trajectories. The HA processing delay Δt_{HA} that would be present in real life (Stone et al., 2008) can be set relative to the loudspeaker-based binaural signal using a variable delay line (VDL). For binaural external sound field reproduction, the loudspeaker signals are pre-processed by applying specifically designed 4-CTC filters, see Section 3.3.3 (Sanches Masiero, 2012). A database of generic or individual playback HRTFs – not necessarily but typically the same as used for the rendering HRTFs – is accessed for the calculation of the CTC filter set. To minimise spectral deviations from an optimally flat loudspeaker frequency response, loudspeaker EQ filters, i.e. inverse on-axis free-field loudspeaker transfer functions, are applied. The listener pose is continuously tracked by means of an optical motion tracking system, see Pausch (2022, Sec. 3.1), and fed back to the HAA module. This pose determines the current sets of spatial transfer functions to be selected from the corresponding databases, as indicated by the dash-dotted feedback lines.

As depicted in the top right, the HA microphone signals, i.e. the HA-related output signals rendered in the HAA module, are re-routed via software loopback to be used as input signals to the MHA and processed by its plug-in chain, including typical HA algorithms. The number of required HA microphone input channels to be processed is determined by the number of channels in the HARTFs. For this specific implementation, two research HAs with two microphones each were used to measure the spatial transfer function datasets.

Combined reproduction in a hearing booth relied on an array of four loudspeakers and a pair of research HAs, as shown in the lower right corner.

Simulation of complex acoustic scenes

Binaural simulation of rooms is based on HRTFs, which enable to efficiently spatialise multiple VSSs given the processing capability of modern computers if only direct propagation paths have to be calculated (Tsingos, Gallo, & Drettakis, 2004), as it is the case in free-field situations. To extend these simulations to include room acoustics, a large amount of room reflections must be spatialised by synthesising BRIRs. Even when using simplified geometric acoustic models, the simulation and synthesis of BRIRs becomes a computationally challenging task when put under real-time constraints (Savioja & Svensson, 2015).

In the current implementation, the real-time framework Room Acoustics for Virtual Environments (RAVEN) was used for this purpose. The simulation models implemented in RAVEN are based on an image source method for combined simulation of the DS and early reflections (Allen & Berkley, 1979; Klein, 2020), while the late reverberation is simulated by means of a ray-tracing algorithm (Krokstad, Strom, & Sørsdal, 1968; Klein, 2020), using a hybrid approach (Schröder, 2011). These simulation components were perceptually optimised to satisfy the requirements of low latency and the limited processing capability of the system. The framework consists of established and perceptually evaluated simulation algorithms (Pelzer, Aretz, & Vorländer, 2011), implemented as C++ libraries, allowing to separately calculate the three perceptually motivated parts of the BRIR.

To account for the simulation of HA microphone signals in addition to BRIRs, simulated at the blocked ear canal entrance, the software implementations were extended to include the HARRIRs, representing the spatial transfer functions at the HA microphone positions. This required to adapt the filter synthesis process to facilitate the use of more than two channels, specifically amounting to four additional channels in the current implementation. It is important to note that the propagation simulation and the filter synthesis are treated as separate rendering processes in the following. In the first process, the calculation of the TOAs and the corresponding levels are calculated for the DS and the early reflections. The filter synthesis subsequently combines the simulation results and merges them with the directivities of the virtual listener and the VSSs of the complex scene.

A reduction of simulation complexity is sometimes permissible, for example, accounting for the fact that head movements in auditory experiments are often limited to head rotations and minor translations only (Lentz et al., 2007). Table 3.1 proposes four possible configurations that allow to calculate perceptually subordinate simulation parts based on pre-calculated databases or during program initialisation instead of during run-time, favouring a decreased number of computations and less computational workload. The configurations always

Table 3.1: Possible configurations for the room acoustic simulations and filter synthesis. The configurations vary with respect to the parts of the BRIR and HARRIR that are either simulated in real-time or on the basis of pre-calculated databases.

Configuration	Direct sound	Early reflections	Late reverberation
A	Real-time updates	Pre-calculated BRIRs/HARRIRs	Pre-calculated BRIRs/HARRIRs
B	Real-time updates	Real-time filter updates, pre-calculated image sources	Pre-calculated BRIRs/HARRIRs
C	Real-time updates	Real-time image source calculation and filter updates	Pre-calculated BRIRs/HARRIRs
D	Full real-time room acoustic simulation and filter updates		

apply to BRIR and HARRIR filter generation. Regarding the computational effort, Configuration A entails the lowest level which gradually increases to Configuration D. Given only limited processing power but a complex scene with multiple VSSs, Configuration A may be a good choice as only the DS is updated in real time, although decreasing the simulation accuracy. Previous work demonstrated that the binaural synthesis of the DS is particularly important (Laitinen et al., 2012; Lindau, 2009), while the directions of arrival of early reflections are only perceivable within a specific duration up to the perceptual mixing time of the BRIRs (Lindau, Kosanke, & Weinzierl, 2012). For entirely static listening experiments, i.e. involving no movements of the listener and VSSs, Configuration A or B provide sufficiently accurate constraints to render the corresponding scene. However, the increase of processing power already on consumer desktop computers makes it possible to address multiple static and moving VSSs using perceptually motivated optimisations. A benchmark analysis, presented in Section 4.3.1, evaluates an example restaurant scene to estimate the achievable filter update rates in case of real-time updates of all BRIR and HARRIR parts, as per Configuration D.

A convenient method to create acoustic scenes for auditory experiments situated in virtual classrooms or restaurants is using 3D computer-aided design software. For this purpose, Aspöck et al. (2014) implemented a plugin for SKETCHUP (Timble Inc., Sunnyvale, California, United States) that allows to position VSSs and a binaural receiver. Room acoustics can be controlled by changing the physical parameters of the surface materials, i.e. absorption and scattering coefficients, to achieve pre-defined target RTs (Pausch et al., 2016c; MacCutcheon et al., 2018; Peng, Pausch, & Fels, 2021).

The real-time software VIRTUAL ACOUSTICS (2021) is also capable of configuring the scene and managing the virtual receiver and the VSSs (Wefers & Vorländer, 2018). Highly optimised algorithms to convolve the simulated BRIRs and HARRIRs with the corresponding anechoic source signals (Wefers, 2015) and a modular concept that allows to render VSSs as configured, for example, only including the DS, represent core features of this software framework. It also addresses Doppler shifts for moving VSSs with changing distance to the receiver (Strauss, 1998; Stienen, 2022). The integrated optical motion tracking system, see Pausch (2022, Sec. 3.1), can be operated at frame rates up to 120 Hz, to capture the current pose of the virtual listener, resulting in simulation updates in the HAA module.

Hearing aid signal processing

Given the extended filter generation module in RAVEN to process HARTF datasets and create the corresponding HARRIRs based on a pre-defined scene, an MHA is integrated for full and transparent control of HA-related signals, that are inserted as pre-processed binaural signals including room acoustic simulations. For the experiments in this thesis, the openMHA (Grimm et al., 2006; Herzke et al., 2017), an open real-time software platform was used, providing established fitting protocols, integrated in an accessible graphical user interface (GUI) for MATLAB™, and a modular concept to combine HA algorithms in a configurable plugin chain with scripting possibilities. These HA algorithms are implemented on the basis of low-latency block-based signal processing concepts.

The two MHA output signals are fed to the direct audio input of the research HAs, see Pausch (2022, Sec. 3.4), to be played back via the left and right HA receiver. Since there is no simultaneous capture and playback by the HA microphones and receivers, respectively, no feedback cancellation is necessary. Nonetheless, it is possible to simulate the feedback path between the acoustic feedback path between the HA receivers and the microphones. With reference to Figure 3.7, this would require an additional convolution per research HA input channel involving the two MHA output signals and the complex transfer function of the feedback path. Finally, the resulting feedback signals are re-routed to the MHA input via software loopback (RME TotalMix, Audio AG, Haimhausen, Germany) and added to the respective MHA input channel.

Combined spatial audio reproduction

Research hearing aids. The used prototypes of research HAs are described in Pausch (2022, Sec. 3.4). These devices, featuring two microphones and one receiver each, were used for all measurements and experiments in this thesis. For acoustic characterisation, the free-field HA microphone transfer functions and SPL receiver transfer functions, as well as the passive damping properties of two different open-fit ear pieces are presented.

External sound field reproduction and reproduction environment. Since the specific implementation of this setup was intended to be used in reproduction environments with only limited space, one 4-loudspeaker array was installed in each of the two hearing booths, cf. Pausch (2022, Sec. 2.1 and 2.2), as described in. Pausch (2022, Sec. 3.5).

System latency

Referring to Table 3.1, it is obvious that the computational workload strongly depends on the selected configuration, simulation parameters and the number of VSSs, making it impossible to derive a single-value EEL to characterise the whole binaural auralisation system. Since the simulation components are separated, listener interaction results in simulation updates executed at different rates, leading to specific latencies per configuration and simulation component. By design, the real-time engine of VIRTUAL ACOUSTICS (2021) processes updates of DS paths after user interaction in the subsequent block of the audio buffer, meaning that the selected buffer size will determine the minimum possible latency. In the current specific implementation, an audio interface (RME Fireface UC, Audio AG, Haimhausen, Germany) with Audio Stream Input/Output (ASIO) protocol was integrated, providing a sufficiently high number of input and output channels, low-latency software loopback, and fast A/D and D/A conversion rates. Operated at a sampling rate of 44.1 kHz, buffer sizes of 128 or 256 samples are a reasonable choice to be set according to the available processing power. The duration it takes the system to react, usually associated with the update of the DS in binaural synthesis, corresponds to the delay of at least one buffer. Note that additional latencies are caused by the motion-tracking system (Friston & Steed, 2014) and the D/A conversion rate of the audio interface, all latency components accumulating to the total EEL.

3.4.2 Combined static hybrid auralisation

Although the proposed combined dynamic binaural auralisation system provides a high level of plausibility, static system implementation variants are the preferred

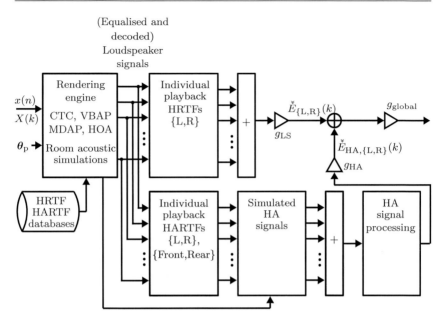

Figure 3.8: Static hybrid HAA system. Schematic depiction of the main elements contributing to the combined perception of the external sound field via the research HAs and the residual ear signals.

choice to systematically evaluate differences across spatial audio reproduction methods and the influence of the reproduction environment. In perceptual evaluations, unless explicitly permitted, unrestricted head movements may add a latent bias or even prevent answering the original research questions owing to noisy empirical data. Objective evaluations based on acoustic measurements that do not address moving VSSs often use an artificial head at a static real-world position, therefore not requiring dynamic or interactive auralisation.

As explained in Section 3.3, loudspeaker-based spatial audio reproduction methods rely on a loudspeaker array surrounding the listener. The two reproduction environments presented in Pausch (2022, Sec. 2.3 and 2.5) contain arrays with $N_{VR} = 16$ and $N_{AC} = 68$ loudspeakers, cf. Pausch (2022, Sec. 3.6) and Pausch, Behler, and Fels (2020), representing possible hardware setups. Figure 3.8 schematically presents the main elements to be accounted for when a listener fitted with research HAs listens to a simulated complex acoustic scene. An input signal $x(n)$ or, equivalently, $X(k)$, is supposed to be rendered and reproduced from a certain direction θ_p via loudspeaker signals, simulated and decoded as necessary for the respective spatial audio reproduction method with optional

loudspeaker EQs. The simulated HA microphone signals, optionally further processed offline or on the MHA, are reproduced via the research HA receivers. In
addition to binaural loudspeaker signals, which are based on databases of HRTFs
or HARTFs, the auralisation framework RAVEN provides simulation modules
to render VBAP and HOA signals including room acoustic simulations (Pelzer,
Masiero, & Vorländer, 2014; Savioja et al., 1999). This allows to assign $x(n)$ to
a VSS that is part of a complex acoustic scene and situated in a virtual indoor
environment, cf. Section 3.4.1, at direction $\boldsymbol{\theta}_p$ and simulate the reflection pattern
at the virtual listener position.

Following the upper reproduction path, the respective method-dependent loudspeaker signals are subsequently convolved with the individual playback HRTFs,
representing a binaural downmix (Jot, Wardle, & Larcher, 1998; Kearney et al.,
2012). As necessary for the respective evaluation approach, cf. Chapters 4 and 5,
these spatial playback transfer functions are time-windowed either in a way that
the acoustic properties of the reproduction environment are preserved, referred
to as playback BRIRs or HARRIRs, or as short as possible to reduce unwanted
reflections, referred to as playback HRIRs or HARIRs. In general, the reflections
from neighbouring loudspeakers, the mounting construction, and a non-ideal reproduction environment will lead to reflections that cannot be avoided even when
applying short time windows. This influence was deliberately neglected in the
previously presented binaural real-time auralisation system assuming negligible
influence of the experimental setup and the reproduction environment. Besides,
the integration of spatial playback transfer functions would require to measure
or simulate filter sets, depending on the reproduction environment and expected
real-world listener poses, on a spherical grid surrounding the listener. Returning
to Figure 3.8, the portion of the external sound field that is perceived through the
blocked ear canal is subsequently summed up per ear, resulting in the binaural
ear signals $\breve{E}_{\{L,R\}}(k)$ or $\breve{E}_{\{L,R\}}(k)$.

In the HA-related path, the microphones of the research HAs will pick up
portions of the external sound field in addition to the simulated HA microphone
signals, represented by the convolution with the individual playback HARTFs
of the left or right, front or rear HA microphones. The resulting by-microphone
signal sets are summed up and used as modified input signals to the MHA to
be optionally processed by HA algorithms. This strategy allows to investigate
the objective and perceptual influence of unwanted reflections in similar environments and hardware setups. Finally, reproduction of the binaural signals via the
HA receivers results in $\breve{E}_{\{L,R\}}(k)$, or $\breve{E}_{\{L,R\}}(k)$.

The two reproduction paths can be weighted individually by the scalar weights
g_{L3} and g_{HA} to allow for relative loudness calibration. To set the combined
reproduction level, a global gain g_{global} is introduced.

4

Objective system evaluation

Objective evaluations of system components and functionalities are crucial to spot and assess strengths and weaknesses in specific implementations. This chapter starts with a description and acoustic analysis of the generic spatial rendering functions, and ideal and in-situ playback transfer functions, which are subsequently used for the objective and perceptual experiments in this thesis. A benchmark analysis of a simulated indoor scene using the specific implementation of the combined binaural real-time auralisation system is presented, allowing to derive recommendations of appropriate system configurations tailored to the processing capabilities of desktop computers. Suitably configured, the combined end-to-end system latency is estimated based on measured IRs. Subsequently, binaural loudspeaker-based reproduction is evaluated for ideal and achievable CS when manipulating various factors, utilising the aforementioned reference and in-situ playback transfer functions. Finally, the influence of the spatial audio reproduction method on recreated receiver transfer functions and binaural cues is analysed by application of the two high-channel loudspeaker arrays.

4.1 Research questions

The research questions listed below shall be answered based on the results of the individual investigations.

Q4.1 What is a suitable system configuration for applications of the binaural real-time auralisation system using a desktop computer?

Related publications:

Pausch et al., 2018b; Pausch, Behler, and Fels, 2020.

Q4.2 Which combined static and dynamic EEL can be expected using the binaural real-time auralisation system?

Q4.3 How does regularisation affect the CS of N′-CTC setups, and what is a
suitable value for β?

Q4.4 Does N′-CTC reproduction in non-ideal reproduction environments enable
to meet the perceptual requirements for CS?

Q4.5 How does the number of loudspeakers N′ affect the in-situ CS of N′-CTC
setups, and which N′ is suitable?

Q4.6 How does the spatial audio reproduction method and the reproduction
environment affect the recreated receiver directivities?

4.2 Generic spatial transfer functions

For the objective evaluations described in this chapter, the directional receiver
characteristics of the adult artificial head without ear simulator, cf. Pausch (2022,
Sec. 3.2), fitted with the pair of research HAs, cf. Pausch (2022, Sec. 3.4), were
acquired either under free-field conditions or at the centre of the spherical loudspeaker arrays in ideal and non-ideal reproduction environments, as described
below. For increased readability, the frequency dependency of the involved filters
and signals is omitted in the following.

4.2.1 Spatial rendering transfer functions with high spatial resolution

Measurement and post-processing

Using the setup described in Pausch (2022, Sec. 3.8), a grid of directions with a
spatial resolution of 1 deg × 1 deg in azimuth and elevation angles was defined.
Accounting for the practical limitation of the zenith angle to 120 deg, two measurement cycles were conducted. Each cycle measured the directions in the upper and
lower hemispheres, covering zenith angles of $0° < \theta \leq 95°$ and $85° < \theta \leq 180°$,
respectively, with the artificial head mounted upside in the second cycle, to be
subsequently combined. The nominal height and viewing direction of the artificial
head in the centre of the measurement sphere was set using a cross line laser.

For the IR measurements, the excitation signal of length 2^{15} samples was
D/A-converted (RME Hammerfall DSP Multiface II, Audio AG, Haimhausen,
Germany), amplified (custom-made amplifier), reproduced via the measurement
loudspeaker, captured by the two microphones at the ear canal entrances and
the front and rear pair of the research HA microphones, A/D-converted (RME
Hammerfall DSP Multiface II, Audio AG, Haimhausen, Germany) and deconvolved. To avoid gain mismatches, each input channel of the audio interface was
calibrated with the aid of a reference voltage source, producing an electrical

1 Vrms sine signal at 1 kHz. After determining the global onset over all directions and subsequently extracting the relevant time segments and removing unwanted reflections from the floor and the measurement setup, the IRs were cropped to a length of 256 samples and time-windowed at the edges (Hann window, 0.25 ms fade in, 2 ms fade out). As reference measurements, the full-sphere measurements were repeated without artificial head present using a free-field microphone (Type 4190 and 2669, Brüel & Kjær, Nærum, Denmark) together with a conditioning amplifier (Type 2610, Brüel & Kjær, Nærum, Denmark), and the two research HAs. All IRs were cropped and time-windowed likewise. The different types of datasets were subsequently convolved with the corresponding inverted reference measurements to obtain the final HRTFs and HARTFs. Device-specific HARTF datasets can be created by referencing all six channels to the measurements of the free-field microphone, thus containing the HA microphone free-field transfer functions, cf. Pausch (2022, Sec. 3.4.2). Remaining level imperfections between the HA microphones in the latter dataset type represent manufacturing variations. To minimise the rendering delay, the contained global delay in both types of 6-channel datasets represents the minimum IR onset determined across all directions of incidence.

Acoustic data analysis

Figures 4.1a and 4.1b present the left-ear magnitude spectra of the HRTFs and front HARTFs, without contained free-field HA microphone characteristics, for directions in the horizontal plane. Right-ear and rear HARTF datasets are not shown as they exhibit similar characteristics. The spectral deviation[1] between the two datasets was obtained by dividing the complex magnitude spectra of the HRTF and the front HARTF datasets, and calculating its magnitude spectrum in dB, cf. Figure 4.1c. Both types of spatial transfer functions are almost uninfluenced by wave interaction effects caused by reflections from the torso and the head below frequencies of approximately 1 kHz. Typical direction-dependent patterns gradually develop with increasing frequency (Møller et al., 1995b). The influence of pinna-related reflections becomes clearly visible at high frequencies above about 7 kHz for ipsilateral directions of incidence, i.e. $0° \leq \varphi \leq 180°$ (Wright, Hebrank, & Wilson, 1974; Blauert, 1997; Raykar, Duraiswami, & Yegnanarayana, 2005). As already hinted in Section 2.8.2, the influence of the pinna will be substantially modified and less pronounced, although not completely absent, when the reception points are located behind the pinna, as it is the case for HARTF datasets. Combined with the spatial offset with respect to the entrance of the ear

---

[1] Not to be confused with the spectral difference (SD) measure, introduced in Section 4.4.2, Equation (4.6).

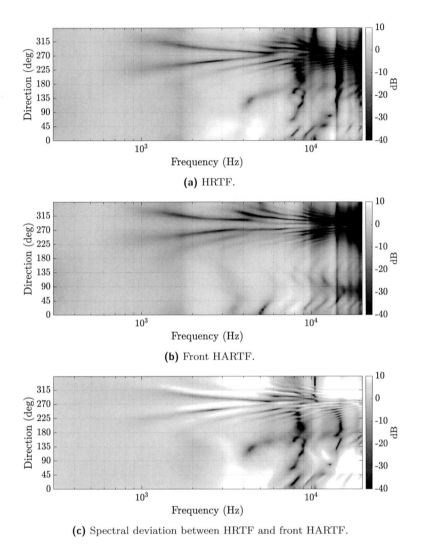

(a) HRTF.

(b) Front HARTF.

(c) Spectral deviation between HRTF and front HARTF.

Figure 4.1: Direction-dependent generic spatial transfer functions in the horizontal plane, $\varphi \in [0°, 359°]$, measured in steps of 1 deg.

canal, leading to additional angular offsets in the resulting spectral patterns, extensive spectral deviations are visible, see Figure 4.1c. For contralateral directions, i.e. $180° < \varphi \le 359°$, combined shadowing and diffraction effects for frequencies above 8 kHz result in spectral patterns gradually developing with frequency and

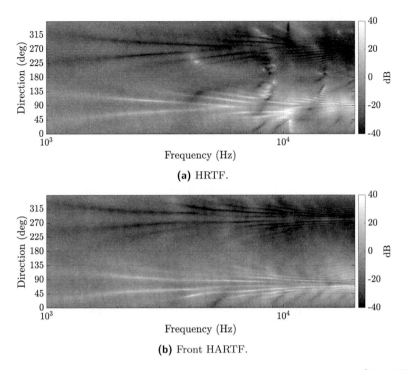

(a) HRTF.

(b) Front HARTF.

Figure 4.2: Direction-dependent generic ILDs in the horizontal plane, $\varphi \in [0°, 359°]$, measured in steps of 1 deg.

direction (horizontal black valleys), as well as narrowband frequency-dependent notches that occur for directional subsets (vertical black valleys between 8 and 11 kHz), cf. Shaw (1974). Particularly the latter effect is not as pronounced and partially absent in the HARTF dataset and thus clearly visible in Figure 4.1c.

Figure 4.2 shows the resulting ILDs of the HRTF and front HARTF datasets for directions in the horizontal plane, cf. Equation (2.16). Although in general quite similar characteristics are present when mirrored around the median plane, deviations stemming from the artificial head's asymmetric anthropometric geometry including different ear geometries are important features that contribute to improved localisation (Bomhardt, 2017). The spectral patterns already observed in Figure 4.1 consequently re-occur in the ILDs. Evaluated between 1.5 kHz and 20 kHz, maximum and minimum ILD values of 64 dB, at $f = 10.9$ kHz and $\varphi = 67°$, and -59 dB, at $f = 19.5$ kHz and $\varphi = 277°$, respectively, can be observed in HRTF data. Similar as for the front HARTF magnitude spectra, a more

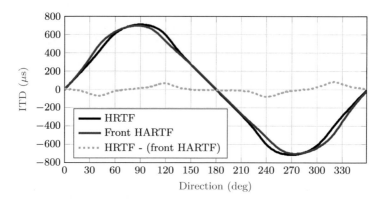

Figure 4.3: Direction-dependent generic ITDs in the horizontal plane, $\varphi \in [0°, 359°]$, measured in steps of 1 deg.

homogeneous spectral pattern emerges in the corresponding ILDs with less disruptive effects of pinna-related reflections. Using the same frequency range for evaluation, lower maximum and minimum ILD values of 45 dB, at $f = 12.4\,\text{kHz}$ and $\varphi = 67°$, and $-56\,\text{dB}$, at $f = 12.7\,\text{kHz}$ and $\varphi = 294°$, respectively, develop. The magnitude reduction of the maxima and minima and the angular shifting of their arguments can be primarily attributed to the horizontal and vertical lateral offsets with respect to the ear canal entrances, reducing the head shadow effect and changing the corresponding maximally effective direction of incidence. Owing to the complex nature of spectral and direction-dependent ILD differences between HRTFs and front HARTFs, general conclusions regarding their perceptual effects cannot be drawn straightforwardly but require further specific investigations.

Compared to the example spatial transfer function, cf. Figure 2.10a, the generic ITDs in the horizontal plane qualitatively exhibits very similar sine-shaped curve progressions and deviations across dataset types, see Figure 4.3. The results are based on the same ITD estimation method with previously applied low pass filter, but evaluated in a frequency range of 100 Hz–1.5 kHz (instead of 500 Hz–1.5 kHz). For reasons already introduced in Section 2.8.2, the HARTF dataset results in ITDs that are angularly shifted and reduced in their maximum magnitudes, compared to the ITDs observed for HRTFs. While the HRTFs entail nearly identical maximum and mimimum ITD values of $\pm712\,\mu s$ at 88 deg and 266 deg, the maximum and minimum ITDs of the front HARTF counterparts amount to 698 μs at 83 deg and $-703\,\mu s$ at 273 deg, respectively. The deviation of maximum and minimum ITD values in HARTF data is directly influenced by the individual placement and the resulting horizontal and vertical lateral offsets of the HAs with respect to the ear canal entrances (Aaronson & Hartmann, 2014). As shown by the blue dotted line, the ITD results mainly deviate in the angular ranges of 0–70 deg (minimum value of $-68\,\mu s$ at 40 deg) and 90–150 deg (maximum value of $68\,\mu s$ at 117 deg), and their symmetric angular equivalents at 210–270 deg (minimum value of $77\,\mu s$ at 239 deg) and 290–360 deg (maximum value of $86\,\mu s$ at 321 deg). Note that substantially larger ITD deviations may be present in individual datasets, cf. Figure 2.10a. These deviations will likely lead to localisation errors in the horizontal plane when utilising HARTFs for binaural localisation. More specifically, negative and positive ITD deviations may lead to an overestimation and underestimation, respectively, of the actual sound source directions in the corresponding horizontal angular quadrants. However, attributing such localisation errors solely to ITD deviations is likely misleading and incomplete (Macpherson & Middlebrooks, 2002).

4.2.2 Spatial playback transfer functions

The influence of the reproduction environment is investigated below based on measured binaural spatial playback between the loudspeakers and the microphones of the artificial head and the research HAs.

Commercial and custom-made hearing booths

For the estimation of the frequency- and rotation-dependent CS in the two hearing booths, cf. Pausch (2022, Sec. 2.1 and 2.2), the artificial head was placed on a turntable and rotated by yaw angles of $\{0°, 20°, 40°\}$ at the nominal listener position and 1.15 m ear height. Each time, the spatial playback transfer functions of the 4-channel loudspeaker arrays, installed at common zenith angles of $\theta = 70°$ and $\theta = 90°$ in the commercial and custom-made hearing booths, respectively, were sequentially measured at the in-ear microphones, driving the loudspeakers with the output signals of the audio interface (RME Fireface UC, Audio AG, Haimhausen, Germany). Note that a rotation of the artificial head only approximates human head rotations since the head-above-torso orientation remains fixed (Brinkmann et al., 2015). Relevant parameters and post-processing steps used for the measurements in the commercial hearing booth are described below (parameters used for the measurements in the custom-made hearing booth are provided in brackets). The excitation signal had a length of 2^{16} samples (2^{17} samples) at $f_s = 48\,\text{kHz}$ ($f_s = 44.1\,\text{kHz}$). Given the flat on-axis frequency responses of the loudspeakers, see Pausch (2022, Fig. 11a), no loudspeaker EQ filters were applied. The raw binaural IRs were linearly convolved with the inverted and time-windowed on-axis in-ear microphone transfer functions (Hann window,

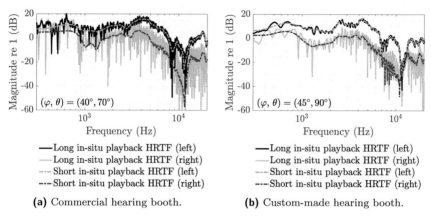

Figure 4.4: In-situ playback transfer functions with long and short time windows for an example loudspeaker direction of the setups in the two hearing booths.

1 ms fade-in after 0 ms, 1 ms fade-out after 11.4 ms), which were implemented as minimum-phase filters. To obtain the short playback HRTFs $\check{H}_{l,\{L,R\}}$, with $l = 1, \ldots, N' = 4$, in each reproduction environment, the left and right sides of Hann windows, each 1 ms long, were subsequently applied, ending 1 ms before and starting 2.2 ms after the IR onset, respectively. After cropping, the IRs were zero-padded to a length of 256 samples. While the same onset windowing was applied to obtain the long in-situ playback HRTFs $\check{H}_{l,\{L,R\}}$, a right-sided Hann window with 5 ms length starting 1 s after the IR onset was used. Note that using such a long time window preserves all room reflections, lying well above the estimated RTs, see Pausch (2022, Fig. 1a). All binaural IRs were globally normalised to 1 in the time domain per rotation-dependent set, without changing relative levels within and across loudspeaker-dependent spatial transfer functions.

The final in-situ playback HRTFs with long and short time windows for the left front loudspeaker are shown in Figure 4.4. Correlating with the RTs, substantially more reflections are visible in the long in-situ playback HRTFs of the commercial hearing booth over the entire frequency range. A high degree of efficiency can be attributed to the acoustic measures in the custom-made hearing booth, resulting in already well-preserved ipsilateral playback HRTFs, even when using a long time window. It is striking that the contralateral ear has a substantially higher reflection density in both experimental environments. This can be traced back to the fact that more reflections emerge off the room surfaces and other array loudspeakers over time, along with longer paths of travels and sound propagation times.

Virtual reality laboratory and anechoic chamber

As was done for the measurements of spatial playback transfer functions in the two hearing booths, the artificial head was placed in a way that the centre of the interaural axis intersects with the centres of the 16-channel and 68-channel loudspeaker arrays, see Pausch (2022, Fig. 12b) and Pausch, Behler, and Fels (2020, Fig. 1). The two experimental environments are referred to as Environment 1 and Environment 2 below. Each time, the target position was set and examined using two self-levelling cross-line lasers (GLL 3-80, Bosch Professional, Gerlingen-Schillerhöhe, Germany) and the optical motion tracking system after translating the head-mounted rigid body to the centre of the interaural axis via offset measurements (Richter, 2019). The final position and orientation errors were within ± 1 mm and ± 2 deg, respectively, in Environment 1, and within ± 3 mm and ± 1.3 deg in Environment 2. All relevant measurement parameters and post-processing steps are presented below, with values in brackets referring to Environment 2.

In-situ playback transfer functions.   The in-situ measurements were conducted at a temperature of 25 °C (22.5 °C) and a relative humidity of 30 % (45 %). The excitation signal had a length of 2^{17} samples (2^{16} samples), both times with $f_s = 44.1$ kHz. For Environment 1, the hardware as used for the room acoustic measurements described in Pausch (2022, Sec. 2.2.3) was applied, except for the reference microphone (Type 4190 and 2669, Brüel & Kjær, Nærum, Denmark) which was operated in combination with a conditioning amplifier (Type 2690, Brüel & Kjær, Nærum, Denmark). The hardware for Environment 2 corresponded to the one described in Pausch, Behler, and Fels (2020) with activated FIR filters. After sequentially measuring the in-situ spatial playback transfer functions between each loudspeaker to the in-ear microphones of the artificial head, and the front and rear pair of the research HAs, the in-situ reference measurements using the reference microphone and the research HAs with absent head were conducted. Since the loudspeakers in Environment 2 were already equalised in time and reproduction levels, no in-situ reference microphone measurements were necessary. Time-windowing was applied to the in-situ reference measurements of the reference microphone and the research HAs, obtained in Environment 1, using two one-sided Hann windows with 1 ms fade-in after 0 ms and 1 ms fade-out after 8 ms (1 ms fade-in after 0.5 ms and 1 ms fade-out after 8 ms) with cropping. The regularised inverse EQ filters were calculated in a frequency range of 90 Hz–20 kHz. To obtain the long and short versions of the playback transfer functions, $\breve{H}$ and $\breve{H}$, the raw binaural IRs were time-windowed using the same time window settings as for the in-situ reference measurements and the right side of a Hann window with 5 ms length beginning at 1 s, respectively, both times involving cropping, to be convolved with the corresponding EQ filters. These referenced IRs were shifted to contain a global delay of 1.5 ms, determined by the earliest IR start in each dataset type, i.e. HRTFs, and front and rear HARTFs combined. Finally, the signals were cropped to a length of 256 samples (short in-situ playback HRTFs and HARTFs) and 44,100 samples (long in-situ playback HRTFs and HARTFs).

Ideal playback transfer functions.   To obtain ideal spatial playback transfer functions without unwanted influence of room reflections, the directions of incidence of the 16-channel and 64-channel loudspeaker arrays were measured in free-field conditions using the setup in the hemi-anechoic chamber, cf. Pausch (2022, Sec. 3.8). Again, the artificial head was mounted upside down to measure directions with $\vartheta > 120°$, as done for the measurements of spatial rendering transfer functions, cf. Section 4.2.1. The same window lengths and post-processing procedure as for the short and long in-situ playback transfer functions was applied. This additional measurement session was necessary to obtain matched directions of incidence that would not have been available with such a high directional precision in the rendering transfer functions, introducing an unwanted source of error.

Results.   The results for the ideal and short in-situ playback transfer functions are presented in Figure 4.5 per loudspeaker array, for example loudspeaker directions corresponding to loudspeakers 10 and 40 in the horizontal plane, cf. Pausch (2022, Fig. 12a) and Pausch, Behler, and Fels (2020, Fig. 2).

In both experimental environments, the associated playback transfer show largely consistent magnitudes up to about 2 kHz, confirming a high directional accuracy across measurement sessions. For higher frequencies, the spectral details start to become progressively blurred due to the influence of room reflections, which results in general spectral curve progression offsets and sometimes less accurate notch depths and quality factors. The high-frequency blurring is slightly more pronounced in the VR laboratory.

To get a better picture of the amount of reflections in each experimental environment, Figure 4.6 presents the post-processed short and long in-situ playback HRTFs for the same example loudspeaker directions. As expected, the VR laboratory exhibits a consistently higher amount of reflections over frequency owing to the reflective room characteristics. Regarding the pronounced reflection pattern in the contralateral ear, similar reasons as provided for the long in-situ playback HRTFs measured in the two hearing booths apply, see Figure 4.4.

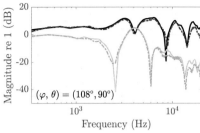

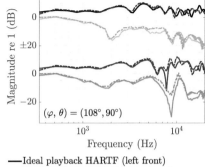

—Ideal playback HRTF (left)
—Ideal playback HRTF (right)
---Short in-situ playback HRTF (left)
--- Short in-situ playback HRTF (right)

—Ideal playback HARTF (left front)
—Ideal playback HARTF (right front)
---Short in-situ playback HARTF (left front)
--- Short in-situ playback HARTF (right front)
—Ideal playback HARTF (left rear)
—Ideal playback HARTF (right rear)
---Short in-situ playback HARTF (left rear)
---Short in-situ playback HARTF (right rear)

(a) virtual reality (VR) laboratory, HRTFs. The legend also applies to sub-figure **(c)**.

(b) VR laboratory, HARTFs. Associated rear HARTFs were shifted by -20 dB. The legend also applies to sub-figure **(d)**.

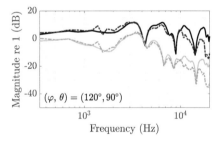

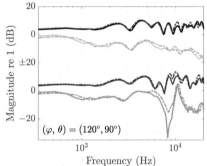

(c) Anechoic chamber, HRTFs.

(d) Anechoic chamber, HARTFs. Associated rear HARTFs were shifted by -20 dB.

Figure 4.5: Ideal, short and long in-situ playback transfer functions for an example loudspeaker direction of the **(a)**–**(b)** 16-channel loudspeaker array in the VR laboratory, and the **(c)**–**(d)** 68-channel loudspeaker array in the anechoic chamber.

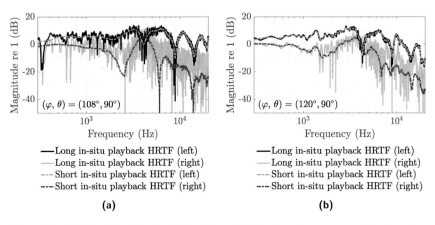

(a)   **(b)**

Figure 4.6: Short and long in-situ playback transfer functions for an example loudspeaker
direction of the **(a)** 16-channel loudspeaker array in the VR laboratory, and
the **(b)** 68-channel loudspeaker array in the anechoic chamber.

4.3 Combined binaural real-time auralisation

4.3.1 Benchmark analysis of a virtual indoor scene

Aiming to find suitable configurations for interactive experiments with accordingly
set simulation parameters, the computational demands of such configurations are
investigated based on a virtual restaurant scene involving multiple VSSs.

As mentioned in Section 3.4.1, it is possible to selectively render only sub-
parts of the BRIRs and HARRIRs in the HAA module. Simulating the DS only
requires performing an audibility test, checking whether the line of sight between
a VSS and the virtual listener is obstructed by objects or walls, and needs to
be updated whenever the VSS or the virtual listener performs a translational
movement. This audibility test is also part of the simulation of early reflections,
during which the image sources and their geometric paths relative to the virtual
listener are calculated and subsequently discarded when inaudible for increased
efficiency. An updated set of image sources has to be accounted for whenever
the positions of the VSS changes, whereas only an audibility test has to be
performed in case of translational receiver movements (Vorländer, 2020). The ray-
tracing algorithm, applied to simulate the late reverberation, relies on frequency-
dependent absorption and scattering coefficients. For the benchmark analysis
below, the audible frequency range is addressed by simulations in octave bands
with centre frequencies between 31.5 Hz and 16 kHz. Setting the length of the
octave band-related energy decay histograms larger than the longest expected RT
of the auralised room ensures valid simulation results.

Example scene

Figure 4.7a shows a screenshot of the 3D model used for the simulations, depicting an example restaurant scene involving three VSSs, with the camera view corresponding to the virtual listener pose. Note that the VSS count includes the two conversations, each one clustered to form one VSS, and the loudspeaker in the upper left corner. To specify the scene more precisely, Figure 4.7b presents its top view, including the pose of the virtual receiver and the positions of the three simulated VSSs. A number of 387 polygons and 109 surfaces construct the 3D model with a room volume of approximately $587\,\mathrm{m}^3$ and a surface area of $625\,\mathrm{m}^2$. Omnidirectional directivities were assigned to all VSSs, making the room acoustic simulation independent of the source signals, which can, for example, be cutlery noise or speech (for S_0 and S_2), or music reproduced by the loudspeaker (for S_1). Such an indoor scene may represent a challenging everyday scenario not only for listeners with NH but particularly for listeners with HL due to interfering noise sources and adverse room acoustical conditions, in combination leading to substantially decreased speech intelligibility, demanding increased listening effort. At the position of the virtual receiver, the mean RT amounts to $T_{30,\mathrm{mid}} = 0.89\,\mathrm{s}$. Utilising the model-based predictor proposed by Lindau, Kosanke, and Weinzierl (2012), a perceptual mixing time of

$$t_{\mathrm{mp95}} = 0.0117\,\frac{\mathrm{ms}}{\mathrm{m}^3} \cdot 581\,\mathrm{m}^3 + 50.1\,\mathrm{ms} \approx 56.9\,\mathrm{ms} \tag{4.1}$$

can be estimated. This duration defines the necessary BRIR filter length which requires updates of the DS and the reflections, corresponding to the perceptual time instance after which it is no longer possible to distinguish individual reflections, cf. Figure 2.3. Thereafter, the late reverberation tail is assumed to be perceptually independent of the virtual listener pose in the simulated room.

Procedure

A two-part benchmark evaluation was performed. While the first part covered the room acoustic simulation for external sound field simulation, the second part included the filter synthesis involving both binaural filters and the HA-related filters. To be able to accurately separate the required calculation times, each part of the simulation was executed one by one. In this way, it is possible to compare the computational demands for each simulation task and avoid biased results owing to multiple threads with different priorities running in parallel, as it is the case in the final system implementation.

The use of one function allowed to collectively evaluate the binaural filters and the HA-related filters, with two and four output channels, respectively, using one synthesis filter with six channels. For both the BRIRs and HARRIRs, a

(a) Screenshot of the underlying 3D model from the perspective of the virtual listener.

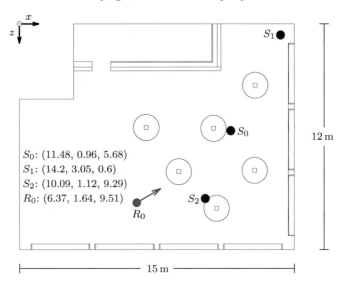

S_0: (11.48, 0.96, 5.68)
S_1: (14.2, 3.05, 0.6)
S_2: (10.09, 1.12, 9.29)
R_0: (6.37, 1.64, 9.51)

(b) Top view showing the pose of the virtual listener R_0 (viewing direction as indicated by the arrow) and the positions of the three VSSs, S_0, S_1 and S_2, all simulated with omnidirectional directivities. The positions are provided as Cartesian coordinate triplets (x, y, z), as per right-handed coordinate system shown in the upper left corner.

Figure 4.7: Complex acoustic scene used for the simulation benchmark analysis involving a virtual listener and three VSSs, set in a virtual restaurant with simulated room acoustics. (Figures taken from Pausch et al., 2018b, licensed under CC BY 4.0, and adapted.)

Table 4.1: Mean calculation times with standard deviations ($\mu \pm \sigma$) and highest possible update rates for the room acoustic simulations. The simulation was separately calculated based on individual models for the DS, early reflections (image sources) and late reverberation (ray-tracing).

Simulation task	VSS			Maximum update rate (Hz)
	S_0 $\mu \pm \sigma$ (ms)	S_1 $\mu \pm \sigma$ (ms)	S_2 $\mu \pm \sigma$ (ms)	
DS, audibility check	$(0.9 \pm 0.0)\text{e}{-}3$	$(3.2 \pm 0.2)\text{e}{-}3$	$(0.3 \pm 0.0)\text{e}{-}3$	75.6e3
Image sources, audibility check	9.9 ± 0.1	13.6 ± 1.0	12.3 ± 4.6	9.3
Image sources, full update	16.9 ± 1.5	16.5 ± 0.3	16.3 ± 0.2	6.7
Ray-tracing (2000 ms)	595.6 ± 8.6	602.5 ± 1.6	577.9 ± 1.8	0.2
Ray-tracing (200 ms)	238.5 ± 1.5	242.8 ± 1.2	224.7 ± 0.9	0.5

filter lenght of 2000 ms was defined. Lying well above the perceptual mixing time, cf. Equation (4.1), an optimised configuration simulating only the first 200 ms of energy decay and reverberation filters with subsequent extrapolation of the remaining filter parts, was simulated. As processing backbone, a computer with an Intel Core i7-3770 CPU @ 3.40 GHz and 64-bit Windows 7 Enterprise operating system was used. All results present average calculation times based on ten simulation cycles. The order of the image sources for the three VSSs was set to 2, and the simulation of the late reverberation relied on 3000 particles per octave bands with activated diffuse rain technique (Schröder, 2011). Since a sequential processing strategy was applied, the maximum update rates were calculated by summing up the calculation times required for all three VSSs and subsequently weighted by 1/3, accounting for simultaneously running tasks. The total CPU time available for the simulation was limited to one third to leave free resources that are required, for example, by the GUI for user/supervisor inputs, the MHA, and the motion tracking system during experiments. Note that the final implementation of the system does not allow accurate control over this settings, owing to varying workload and multithreading (Aspöck et al., 2014).

Results

Table 4.1 presents the mean results split into five different simulation tasks. Fastest update rates of about 76 kHz are possible for the DS. For the early

Table 4.2: Mean calculation times with standard deviations ($\mu \pm \sigma$) and highest possible update rates for the filter synthesis of the DS, early reflections (image sources) and late reverberation (ray-tracing). The combined filters, based on HRIRs and HARIRs with 128 samples length and $f_s = 44.1\,\mathrm{kHz}$, were calculated for six channels (2 binaural channels, 4 HA-related channels).

	VSS			Maximum
	S_0	S_1	S_2	update
Filter synthesis task	$\mu \pm \sigma$ (ms)	$\mu \pm \sigma$ (ms)	$\mu \pm \sigma$ (ms)	rate (Hz)
DS	0.1 ± 0.0	0.1 ± 0.0	0.1 ± 0.0	923.7
Image sources	0.2 ± 0.0	0.2 ± 0.0	0.2 ± 0.0	545.6
Ray-tracing (2000 ms)	196.3 ± 0.7	195.2 ± 1.1	196.1 ± 1.0	0.6
Ray-tracing (200 ms)	19.6 ± 0.1	19.6 ± 0.1	20.4 ± 0.3	5.6

reflections, the simulation of which is based on about 22 audible image sources per VSS, these updates rates lie within a range of 6–10 Hz. A further increase of calculation times has to be accepted for the simulation of the late reverberation using ray-tracing. Accounting for the room geometry and calculating the energy decay correspondingly for each VSS, update rates of 0.2 Hz were obtained, which improved to 0.5 Hz in the optimised filter configuration.

In Table 4.2, the filter synthesis results for the individual parts of the room IR are presented. High filter synthesis update rates of about 924 Hz for the DS favour a plausible perception, since this part of the IR is the main contributor to VSS localisation. Although substantially dropping to about 546 Hz, the filter update rates for the image sources are still above 100 Hz and can be considered as sufficiently high (Vorländer, 2020; Lentz, 2008). Lowest update rates of 0.6 Hz were again measured for the late reverberation with long filter, with improved results for the optimised filter length, allowing 5.6 Hz.

Discussion

Both the simulation and filter synthesis benchmark analysis confirmed that the auralisation engine is capable of providing very high update rates of the DS. In interactive auralisations and dynamic scenes, involving a moving receiver and VSSs, related audibility checks for the DS can be updated equally fast. Since it is required to generate a filter for every sample buffer per VSS, the filter synthesis of DS represents a crucial operation. Collectively, the results on the DS indicate that the HAA module fulfils key criteria towards plausibility.

When it comes to the simulation and filter synthesis of early reflections and late reverberation, the resource limits of the used computer became visible, restricting the number of possible VSSs and the allowed complexity of room models, since otherwise the simulation may deliver potentially invalid or perceptually inadequate results. Remedy is provided, for example, by decreasing the image source order to 1, which may still be enough to deliver plausible presentation of early reflections, depending on the specific experimental requirements.

Selecting Configuration A, see Table 3.1, represents the most conservative choice, since only the DS needs to be updated, while accessing pre-calculated databases of BRIRs and HARRIRs for the simulation of reflections. Applying the parameter settings presented in Table 4.3, which were also used for the EEL latency measurements in the upcoming section, the binaural HAA system delivered faultless auralisation results of the example scene without any noticeable artefacts in Configuration A. For the three VSSs, 12 convolutions for the room acoustic filters and 12 convolutions for the DS were necessary. However, increasing the length of the room acoustic filters and the number of VSSs, elevated the probability of audible glitches and dropouts, particularly in situations with simultaneous workload, caused, for example, by the MHA or the MATLAB™ GUI used for collecting the empirical data during the experiment. Further specific investigations and optimisations are necessary to improve the system performance using consumer hardware with limited processing power.

The fact that the late reverberation in conventional rooms at a listener position does not vary substantially relaxes somewhat the need for fast simulation updates. Behavioural results from listening experiments (Lentz, 2008), in which listeners were encouraged to interact with the virtual scene, further revealed that most users only applied translational and rotational movements with velocities of less than 16 cm/s and 10 deg/s. In rather static scenes with simple geometry, a possible strategy would thus be to pre-calculate echograms and reverberation filters and only exchange them when exceeding certain motion thresholds (Aspöck et al., 2014). While this simplification will not serve physical correctness, it still allows for plausibility and increased efficiency when recreating interactive virtual scenes.

To answer research question Q4.1 based on the benchmark results, Configuration A, B, or C, cf. Table 3.1, are the recommended configurations for desktop computers, which should be preferred over full real-time simulation, i.e. Configuration D.

4.3.2 Combined system latency

The definition of EEL in the proposed system relies on the absolute TOAs of the respective IRs in each of the two auralisation paths, accounting for the delays during rendering and reproduction. Since, to my knowledge, there is no publication about the latency of the integrated optical tracking system using the exact same hardware and software (Flex 13; Motive 1.8.0 Final; NaturalPoint, Inc. DBA OptiTrack, Corvallis, OR, USA), only the static EEL was measured,

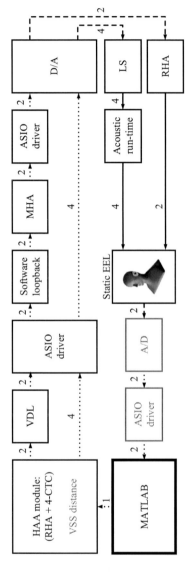

Figure 4.8: Block diagram showing the signal flow for measuring static EEL with main contribution elements. To measure the path-dependent IRs, a VSS reproduced the measurement signal in a virtual room. The HAA module accessed databases of pre-calculated filters to simulate the HA-based signals, which were delayed using a VDL and processed on the MHA, and loudspeaker-based signals. Each signal path was reproduced via a set of four loudspeakers applying 4-CTC filter network, and the two research HA receivers, captured by the artificial and post-processed in MATLAB™. Signals in the digital and analogue domains are indicated by dotted and dashed lines, respectively, and acoustic signals are represented by solid lines. (Figure taken from Pausch et al., 2018b, licensed under CC BY 4.0.)

while estimating the tracker latency, and the resulting dynamic EEL, based on results obtained with similar hardware used in previous work.

Procedure

For the measurement, the same artificial head as used for the acquisition of generic spatial transfer functions, cf. Section 4.2, was placed in the centre of the commercial hearing booth, cf. Pausch (2022, Sec. 2.1), at an ear height of 1.2 m, wearing the two research HAs. To accurately estimate the real-world pose of the artificial head relative to the global loudspeaker positions, a rigid body was attached to the head, correcting the resulting offset of its pivot point to the centre of the interaural axis (Pausch et al., 2016a; Pausch & Fels, 2020; Richter, 2019). In Figure 4.8, various system components that contribute to the static EEL are shown, including the signal flow.

The virtual indoor scene definition included a VSS, placed in the horizontal plane at $\varphi = 0°$ at a distance of 2 m. An exponential swept sine with a length of 2^{16} samples was reproduced by the VSS. A desktop computer (Intel Core i7-4770 @ 3.4 GHz, Windows 7 Enterprise) was used to synthesise all room acoustic filters, based on BRIRs and HARRIRs, in the HAA module, corresponding to Configuration A in Table 3.1. The VDL in the HA-based path was configured to produce a delay of 5 ms relative to loudspeaker-based reproduction (Stone et al., 2008). These signals were re-routed via software loopback (RME TotalMix, Audio AG, Haimhausen, Germany) to be used as input signals for the MHA Grimm et al., 2006, whose plug-in chain consisted of a downsampling stage (factor 2, plug-in `downsample`), a calibration stage (plug-in `splcalib`), accessing sub-plug-ins such as a Fourier transform filterbank, a limiter, compressor, and an overlap-add plug-in for re-synthesis purposes. Subsequent upsampling (factor 2, plug-in `upsample`), D/A conversion and amplification in the audio interface conditioned the signals for playback via the research HA receivers. Similarly, the loudspeaker-based signals, prepared in the HAA module as per 4-CTC approach, were D/A-converted, amplified and reproduced via the 4-channel loudspeaker array, see Pausch (2022, Sec. 3.5), including the acoustic run-time between the loudspeakers and in-ear microphones of the artificial head. The two in-ear microphones of the artificial head measured the path-dependent sweep responses, which were A/D-converted and amplified by the same audio interface, and deconvolved in MATLAB™.

Indicated by grey font and boxes in Figure 4.8, the VSS distance and the delay of the A/D conversion and the ASIO driver were subtracted, since these components are not contributing to the final static EEL when using the system. Table 4.3 presents the specific parameter settings of system components as configured for the described combined latency measurements.

Table 4.3: Specific system configuration for static EEL measurements. Settings related to the MHA only show the changes that have been applied to the provided standard configuration file `mha_hearingaid.cfg`.

Component	Setting
Audio interface	
Sampling rate	44100 Hz
Buffer size	128 samples
Measurement signal	
Frequency range	20–20000 Hz
Length	2^{16} samples
HAA module	
Number of channels	6
HRIR filter length	256 samples
HARIR filter length	256 samples
BRIR filter length	44,100 samples
HARRIR filter length	44,100 samples
CTC filter length	1226 samples
Regularisation parameter β	0.01
MHA	
Number of channels	2
Sampling rate	22050 Hz (downsampled)
Fragment size	256 samples
Plug-in chain	`[downsample ...`
	`splcalib upsample]`

Results

The additional delay caused by the acoustic runtime of the VSS to be removed corresponded to 256 samples, or 5.8 ms, at a speed of sound of 344 m/s. An input latency of 148 samples, or 3.4 ms, was displayed by the ASIO driver interface (RME Fireface UC, driver version 1.099, hardware revision: 133) for the selected buffer size and likewise removed. Corrected by the unwanted delay components, the left-ear IRs are displayed in Figure 4.9, featuring delays of $\Delta t_{\mathrm{CTC}} = 1142$ samples (25.9 ms) and $\Delta t_{\mathrm{HA,absolute}} = 1363$ samples (30.9 ms) for the loudspeaker-based signal and the research HA-related path, respectively. The pre-defined relative delay of $\Delta t_{\mathrm{HA}} = 5$ ms between the two paths was confirmed by the measurement results.

Designed for interactive listening scenarios, the dynamic EEL of the auralisation system is of particular interest and will be estimated based on measurement results from the literature. Teather et al. (2009) reported latency values of 73 ± 4 ms utilising similar optical motion tracking systems (Flex:C120; 120 Hz frame rate; NaturalPoint, Inc. DBA OptiTrack, Corvallis, OR, USA). Upon request, Friston and Steed (2014) disclosed having used very similar hardware

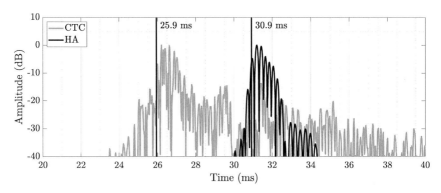

Figure 4.9: Measured path-related IRs, corrected by unwanted delay components, their onsets defining the static EEL. Note that only the left-ear signals of the artificial head are shown. (Figure taken from Pausch et al., 2018b, licensed under CC BY 4.0, and adapted.)

from the same manufacturer but with lower image resolution and native frame rate (Flex 3; 100 Hz frame rate; NaturalPoint, Inc. DBA OptiTrack, Corvallis, OR, USA) for their investigations. On average, they measured latency values of 50.43 ms (with maximum values of $\Delta t_{\text{tracking,max}} = 54\,\text{ms}$) for the configuration "PC 3 OptiTrack Motive Rigid Body Aero Off" on a system with Windows 7. With the help of the latter values, it is possible to estimate an upper dynamic EEL value by adding the maximum latency to the measured static EEL for the loudspeaker-based signals, yielding

$$\begin{aligned} \text{EEL}_{\text{dynamic,CTC}} &= \text{EEL}_{\text{static,CTC}} + \Delta t_{\text{tracking,max}} = \\ &= 25.9\,\text{ms} + 54\,\text{ms} = 79.9\,\text{ms}, \end{aligned} \tag{4.2}$$

which answers research question Q4.2.

Discussion

Following up on the requirements on system latency, introduced in Section 3.2.6, the measured static EEL falls below the perceptual detection thresholds. Vorländer, 2020 further suggests a general maximum latency of 50 ms for VAEs, which is also upheld by the static EEL results. In view of the compliance with the relative delay $\Delta t_{\text{HA}} = 5\,\text{ms}$, it was shown that the system is capable of processing and reproducing the simulated signals under real-time constraints, while still allowing to control the temporal delay of the HA-related signals as required to replicate processing delays observed in the real-world equivalent of the simulated acoustic scene involving HAs.

However, when integrating motion tracking to make the system fit for the purpose originally intended, the estimated dynamic EEL slightly exceeds the detectable and recommended threshold values of 50 to 75 ms (Brungart, Simpson, & Kordik, 2005; Yairi, Iwaya, & Suzuki, 2006; Vorländer, 2020) but lies well below 107.63 ± 30.39 ms, the threshold reported by Lindau (2009). Since the optical tracking system used for the specific implementation exhibits a slightly higher native frame rate of 120 Hz, the added delay $\Delta t_{\mathrm{tracking,max}}$ represents a conservative upper estimate. Informal perceptual tests further attribute a high responsiveness to the implemented system, which is why the determined dynamic EEL was subjectively classified as acceptable for the planned clinical experiment, cf. Chapter 6.

4.4 Combined static hybrid auralisation

4.4.1 Channel separation in binaural loudspeaker reproduction

In binaural reproduction systems over loudspeakers, the CS represents a widely used metric to investigate the performance of CTC systems by assessing its ability to deliver and suppress the binaural signals at the target ear and the opposite ear, respectively. Simulating a binaural signal with content for the left (target) ear only requires setting the binaural input signal to $\mathbf{b} = [1, 0]^{\mathsf{T}}$ (Sanches Masiero, 2012) and substituting it into Equation (3.1), which leads to the ear signals[2]

$$E_{\mathrm{CTC,L}} = \breve{H}_{1,\mathrm{L}} C_{\mathrm{L},1} + \breve{H}_{2,\mathrm{L}} C_{\mathrm{L},2} + \cdots + \breve{H}_{\mathrm{N'},\mathrm{L}} C_{\mathrm{L},\mathrm{N'}} \tag{4.3}$$

$$E_{\mathrm{CTC,R}} = \breve{H}_{1,\mathrm{R}} C_{\mathrm{L},1} + \breve{H}_{2,\mathrm{R}} C_{\mathrm{L},2} + \cdots + \breve{H}_{\mathrm{N'},\mathrm{R}} C_{\mathrm{L},\mathrm{N'}}. \tag{4.4}$$

The resulting CS for the left ear is thus defined as the logarithmic magnitude ratio

$$\mathrm{CS_L} = 20 \log \left(\frac{|\breve{H}_{1,\mathrm{L}} C_{\mathrm{L},1} + \breve{H}_{2,\mathrm{L}} C_{\mathrm{L},2} + \cdots + \breve{H}_{\mathrm{N'},\mathrm{L}} C_{\mathrm{L},\mathrm{N'}}|}{|\breve{H}_{1,\mathrm{R}} C_{\mathrm{L},1} + \breve{H}_{2,\mathrm{R}} C_{\mathrm{L},2} + \cdots + \breve{H}_{\mathrm{N'},\mathrm{R}} C_{\mathrm{L},\mathrm{N'}}|} \right). \tag{4.5}$$

Conversely, setting $\mathbf{b} = [0, 1]^{\mathsf{T}}$ allows to assess the right-ear CS. In a well-designed CTC system, the CS is superior to the one that would occur only due to effects of the head shadow, referred to as natural CS (Blauert, 1997; Sanches Masiero, 2012). This is particularly true for the low frequency range where head shadow effects are less pronounced or completely absent. As per definition, a higher CS suggests

---

[2]Note that $\breve{H}$ may represent playback HRTFs that were selected from the database of rendering HRTFs with high spatial resolution, cf. Sections 4.2.1 and 3.3.3, or short playback HRTFs, cf. Section 4.2.2, as required for the evaluation scenarios described below.

better effectiveness and thus overall performance of the CTC system (Akeroyd et al., 2007; Bai & Lee, 2006).

Parodi and Rubak (2011) manipulated the amount of CS in a virtual 2-channel CTC setup, with simulated CTC-filtered binaural signals reproduced via headphones, for various stimuli, loudspeaker span angles and listener position offsets. The reported thresholds correspond to the minimum CS that is required that a binaural signal with CS cannot be distinguished from a binaural signal without acoustic crosstalk. For speech stimuli, broadband and narrowband noise with centre frequencies lower than 1 kHz, their experimental results suggested required average values of 20 dB given that the listener is located at the nominal listener position. Slightly decreased thresholds of 15 dB were reported if the centre frequencies of the narrowband noise stimuli fall between 1 and 2 kHz. Rooting in the sensitivity to ILD modifications, required thresholds as high as 25 dB were found for higher frequency bands. While correlations between sagittal plane localisation performance were reported to be weak, CS can still be considered as a useful measure influencing localisation in the horizontal plane (Sanches Masiero, 2012; Majdak, Masiero, & Fels, 2013).

Effect of regularisation

An example 4-channel loudspeaker array, as installed in the custom-made hearing booth, see Pausch (2022, Sec. 3.5), was simulated. The generic rendering HRTFs, cf. Section 4.2.1, for the corresponding loudspeaker azimuth directions with common zenith angle $\theta = 70°$, were selected after applying a nearest neighbour search, given a nominal viewing direction of $(\varphi, \theta) = (0°, 90°)$ of the listener. Note that the subset of HRIRs was globally normalised to 1 in time domain, while preserving all relative level differences within and across directions. These normalised spatial transfer functions were used as playback HRTFs to calculate the 4-CTC filters of 2^{12} samples length with different values for β, cf. Equation (3.5), and estimate the highest possible optimal CS.

The frequency-dependent results are shown in Figure 4.10. Largest optimal broadband CS values of 73.7 ± 5.3 dB ($\mu \pm \sigma$), evaluated for 100 Hz–20 kHz, with a tendency to increase towards higher frequencies, are achieved using no regularisation, i.e. $\beta = 0$, which in turn leads to unpleasantly sharp peaks and strong colouration artefacts, see Figure 3.5a. Increasing the amount of regularisation has no noticeable effect at first, yielding 73.7 ± 6.0 dB ($\mu \pm \sigma$, $\beta = 0.001$) but gradually lowers the optimal broadband CS to 62.9 ± 10.8 dB ($\mu \pm \sigma$, $\beta = 0.01$), 52.3 ± 14.4 dB ($\mu \pm \sigma$, $\beta = 0.05$), and 47.3 ± 15.2 dB ($\mu \pm \sigma$, $\beta = 0.1$). With increasing regularisation, the monaural cues in playback HRTFs can no longer be addressed in the same degree of detail, which entails local performance drops

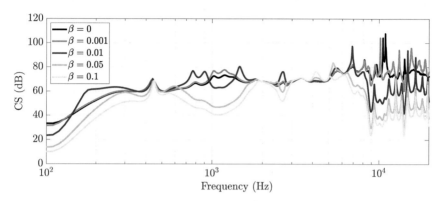

Figure 4.10: Optimal CS with varying regularisation β for a simulated 4-CTC loud-speaker arrangement.

starting at frequencies above about 8 kHz for the highest three regularisation values. Additional reductions in CS can be observed towards frequencies below $f < 300$ Hz and between 600 Hz and 1.9 kHz for $\beta = \{0.05, 0.1\}$. Although the averaged broadband values would suggest perceptually sufficient CS, it is recommended not to apply regularisation factors higher than $\beta = 0.01$, since the achievable CS will be further reduced in actual experimental environments, as demonstrated in the upcoming investigations. In the optimal scenario, the use of this maximum recommended regularisation allows to exceed the perceptual broadband threshold of 20 dB (Parodi & Rubak, 2011) towards frequencies below 200 Hz while reducing unwanted CTC filter peaks to a perceptually acceptable extent, which was confirmed by informal perceptual tests, and is therefore used for all evaluations presented below.

Regarding research question Q4.3, it was shown that increasingly high regularisation β progressively deteriorates CS, particularly towards high frequencies. The use of a recommended β of 0.01 allows to decrease unpleasant filter characteristics while still fulfilling perceptual requirements on minimum CS.

Effect of the reproduction environment

Figure 4.11 and Table 4.4 present the optimal and achievable left-ear in-situ CS in the commercial and custom-made hearing booths after applying the CTC filters, calculated based on $\breve{H}_{\{l\},\{L,R\}}$, see Section 4.2.2 and Equation (4.5), on $\breve{H}_{\{l\},\{L,R\}}$ and $\breve{H}_{\{l\},\{L,R\}}$. Such an approach allows to estimate the optimal and achievable in-situ CS. The broadband results represent CS values averaged across a frequency range of 100 Hz–20 kHz. Note that the optimal in-situ CS differs from

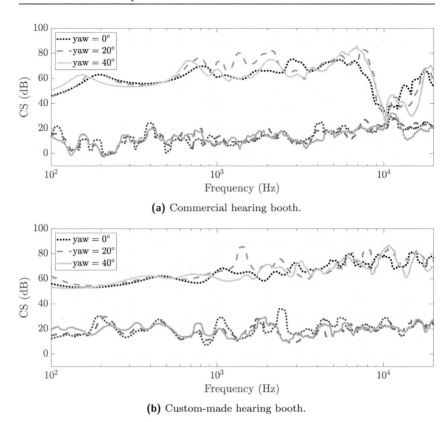

(a) Commercial hearing booth.

(b) Custom-made hearing booth.

Figure 4.11: Optimal and achievable in-situ CS, plotted in grey and blue colour grada-
tions, respectively, using the 4-CTC ($\beta = 0.01$) loudspeaker setups installed
in the hearing booths. The CS performance was calculated for selected yaw
angles of the artificial head placed at the nominal array centre, as indicated
by the corresponding line types.

the one evaluated in the paragraph above, since even applying a short time window
will not allow to exactly obtain the corresponding ideal spatial playback transfer
functions owing to reflections off neighbouring room surfaces and loudspeakers,
see Section 4.2.2.

Optimal in-situ CS values are largely above the achievable in-situ CS except around 10 kHz in the commercial hearing booth, and generally exhibit substantially higher values. The mentioned performance drop for frequencies above
around 7 kHz can be traced back to effects of regularisation, hindering to address
more complex combinations of spatial playback transfer functions $\check{H}_{\{l\},\{L,R\}}$

Table 4.4: Optimal and achievable in-situ CS in the commercial and custom-made hearing booths, evaluated for different frequency ranges and yaw angles of the artificial head.

In-situ CS type	Hearing booth	Yaw angle (deg)	Frequency range: Broadband $\mu \pm \sigma$ (dB)	0.1–1 kHz $\mu \pm \sigma$ (dB)	2–16 kHz $\mu \pm \sigma$ (dB)
Optimal	Commercial	0	51 ± 17.2	61 ± 5.6	54.5 ± 16.1
		20	52.5 ± 19.0	62.8 ± 7.9	53.8 ± 18.3
		40	51.5 ± 19.1	61.7 ± 6.9	53.2 ± 19.7
	Custom-made	0	67.8 ± 12.8	58.4 ± 3.2	71.3 ± 6.1
		20	69.2 ± 14.0	60.2 ± 3.0	73 ± 6.1
		40	66.5 ± 14.6	60.4 ± 3.0	70.5 ± 8.6
Achievable	Commercial	0	17 ± 7.1	7.8 ± 7.1	17 ± 7
		20	17.6 ± 7.1	8.8 ± 7.2	17.7 ± 7.5
		40	16.1 ± 7.7	9.8 ± 6.8	15.8 ± 8.1
	Custom-made	0	20.2 ± 6.2	16.2 ± 5.9	20 ± 6
		20	19.1 ± 5.9	17 ± 5.2	18.8 ± 5.5
		40	19.3 ± 6.1	18.7 ± 4.1	18.5 ± 5.7

and resulting spectral details, which decreases for values of $\beta < 0.01$ and even vanishes in case of applying no regularisation. More consistent optimal in-situ CS results with no performance drop in the mentioned frequency region were achieved in the custom-made hearing booth, already hinting increased effectiveness of the room acoustic optimisation measures. In comparison, the broadband and high-frequency results are consistently higher in the custom-made hearing booth, while both reproduction environments exhibit comparable optimal results in the low-mid frequency range.

Rotating the artificial head yielded similar results per in-situ CS type and hearing booth. This suggests that the N′-CTC filter network is capable of providing effective filtering in case of user movements, allowing interactive auralisation of complex acoustic scenes. In this regard, the CS results are transferable to the specific implementation of the binaural real-time auralisation system, cf. Section 3.4.1.

The drop in the achievable in-situ CS reveals the dominant effect of reflections on the efficacy of CTC filters in non-ideal reproduction environments (Sæbø, 2001). Kohnen et al. (2016) extended the CTC filters to include second-order reflections and demonstrated improved CS performance in simulations relying on geometrically simple rooms with reflective surfaces. However, this performance increase could not be observed when including measured in-situ playback transfer functions with complex reflection patterns due to complex acoustic effects, such as scattering and diffraction, that were not addressed by the simulation. Such effects are also present in the current experimental environments. Averaged across yaw angles, the reduction in CS amounted to -36.8 ± 2.9 dB ($\mu \pm \sigma$, broadband), -52.5 ± 1.0 dB ($\mu \pm \sigma$, 0.1–1 kHz), and -38.5 ± 2.2 dB ($\mu \pm \sigma$, 2–16 kHz) in the commercial hearing booth. In the custom-made hearing booth, performance drops of -48.3 ± 1.5 dB ($\mu \pm \sigma$, broadband), -42.4 ± 0.8 dB ($\mu \pm \sigma$, 0.1–1 kHz), and -52.5 ± 1.5 dB ($\mu \pm \sigma$, 2–16 kHz) were observed. Regarding the absolute achievable in-situ CS, the results in the commercial hearing booth are consistently below the recommended minimum CS values, while the values obtained in the custom-made hearing booth on average are very close or above the requirements for broadband stimuli and narrowband noise with centre frequencies below 1 kHz, respectively, falling below the elevated minimum thresholds in the high-frequency range (Parodi & Rubak, 2011). It should be noted, however, that this assessment of CS represents a rigorous approach and comparison since the reported minimum CS values were based on perceptual evaluations using simulated variants of CTC systems without additional room reflections. Due to the fact that humans primarily evaluate the DS when localising VSSs, calculating the CS including early and late reflections may lead to unduly stringent conclusions from a perceptual perspective. This said, the perceptively exploitable CS probably lies between the achievable in-situ CS, representing the worst case scenario, and the optimal in-situ CS.

To follow up on research question Q4.4, the perceptual demands in terms of achievable in-situ CS are in a critical range in the commercial hearing booth, and is only met by the custom-made hearing booth. The perceptual results corroborate a potential negative effect on localisation performance in the former experimental environment, as discussed in Section 5.1.

Effect of the number of involved loudspeakers

As explained in Section 3.3.3, it is theoretically possible to use all N loudspeakers of a loudspeaker array surrounding the listener given enough processing capacity. However, for real-time applications, it is of interest if a number of loudspeakers $N' < N$ is sufficient to obtain a perceptually high enough CS. In this regard, it also shall be investigated how an increasing number N' affects the CS. Therefore, different subset layouts with double the number of involved loudspeakers compared to the previous layout, including the full layouts, were selected for the evaluations, see Figure 4.12. For $N' \leq 4$, the layouts relied on elevated loudspeaker directions (Lentz, 2008; Parodi & Rubak, 2010; Sanches Masiero, 2012), whereas layouts with higher loudspeaker counts aimed at balanced spatial distributions. Applying the CTC filters, calculated for the subset loudspeaker layouts

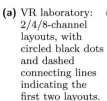

(a) VR laboratory: **(b)** Anechoic   **(c)** Anechoic   **(d)** Anechoic
2/4/8-channel   chamber: chamber: chamber:
layouts, with   2/4/8-channel   16-channel layout.   34-channel layout.
circled black dots   layouts, with
and dashed   circled black dots
connecting lines   and dashed
indicating the   connecting lines
first two layouts.   indicating the
                       first two layouts.

Figure 4.12: Subset layouts of the loudspeaker arrays **(a)** in the VR laboratory, and **(b)**–
(d) in the anechoic chamber, not showing the corresponding full layouts, as
used for the estimation of optimal and achievable in-situ CS. The nominal
listener pose is indicated by view and up vectors.

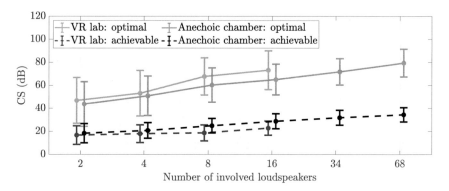
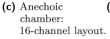

Figure 4.13: Optimal and achievable in-situ CS with varying number of involved loudspeakers selected as per the loudspeaker subset scenarios presented in Figure 4.12, and the full set of loudspeakers per array.

based on the equalised short in-situ playback HRTFs, cf. Section 4.2.2, and on
the corresponding short and long in-situ playback HRTFs, the broadband and
frequency range-specific optimal and achievable left-ear in-situ CS was estimated,
see Figure 4.13 and Table 4.5.

As already discussed earlier, a substantial performance drop between optimal
and achievable in-situ CS values can be reported, on average amounting to
-41.2 ± 10.1 dB ($\mu \pm \sigma$, VR laboratory) and -35.4 ± 7.0 dB ($\mu \pm \sigma$, anechoic
chamber). Reflections from room surfaces (VR laboratory) in addition to those

Table 4.5: Optimal and achievable in-situ CS in the VR laboratory and anechoic
chamber, evaluated for different frequency ranges and number of involved
loudspeakers N′.

In-situ CS type	Reproduction environment	N′	Frequency range		
			Broadband $\mu \pm \sigma$ (dB)	0.1–1 kHz $\mu \pm \sigma$ (dB)	2–16 kHz $\mu \pm \sigma$ (dB)
Optimal	VR laboratory	2	46.8 ± 19.9	41.2 ± 12.8	47.8 ± 21.2
		4	53 ± 19.8	51.3 ± 10.6	55.6 ± 19.9
		8	67.6 ± 16.1	60.7 ± 9.8	71.1 ± 12.1
		16	73.1 ± 16.8	64.2 ± 9.4	76.7 ± 9.0
	Anechoic chamber	2	43.8 ± 19.4	44.3 ± 7.2	47.2 ± 21.2
		4	50.9 ± 17.2	51.4 ± 7.6	56.6 ± 16.1
		8	60.3 ± 14.9	59.8 ± 9.6	65.8 ± 11.6
		16	65.1 ± 13.3	62.2 ± 8.9	70.7 ± 9.2
		34	71.8 ± 11.5	63.7 ± 6.3	75.5 ± 7.9
		68	79.5 ± 12.0	65.3 ± 4.8	84.3 ± 6.7
Achievable	VR laboratory	2	16.7 ± 8.1	8.1 ± 6.7	16 ± 7.3
		4	17.9 ± 7.7	10.4 ± 5.2	17.4 ± 7.4
		8	18.6 ± 6.9	11.2 ± 5.6	18.2 ± 7.2
		16	22.7 ± 6.1	15.9 ± 6.0	22.5 ± 5.7
	Anechoic chamber	2	18.4 ± 8.3	23.6 ± 5.1	17 ± 9.1
		4	20.8 ± 6.8	25.1 ± 5.8	20.8 ± 7.3
		8	24.9 ± 6.3	27.5 ± 5.9	24.4 ± 6.1
		16	28.8 ± 6.5	28 ± 5.1	28.4 ± 5.8
		34	31.9 ± 6.5	29.8 ± 6.2	31.1 ± 5.7
		68	34.5 ± 6.2	30 ± 6.8	34.6 ± 5.9

from the mounting construction and neighbouring loudspeakers (VR laboratory
and anechoic chamber) in combination with the larger array radii, exhibiting
higher reflection potential in the playback HRTFs, already affect the optimal
in-situ CS. While the optimal in-situ CS for N′ = 4 is similar to the one achieved
in the commercial hearing booth, the performance is consistently lower compared
to the results obtained in the custom-made hearing booth, cf. Tables 4.4 and 4.5.
This performance difference again confirms the need for acoustically optimised
reproduction environments, a compact loudspeaker design and an unobtrusive
array mounting construction, particularly when using hardware setups with a
high loudspeaker count, even when set up in an anechoic chamber.

Each doubling of N′ amounted to comparable mean increases of 8.8 ± 5.1 dB
($\mu \pm \sigma$) and 6.5 ± 1.6 dB ($\mu \pm \sigma$) regarding the optimal in-situ CS for the setups
in the VR laboratory and the anechoic chamber, respectively. In the first setup,
the stepwise increase was between 5.4 dB and 14.6 dB, with the highest jump
from N′ = 4 and N′ = 8, while the performance improved more constantly across

subset selections in the second setup, ranging from 4.8 dB to 9.4 dB. On average, the achievable in-situ CS was consistently lower in the first and second setup, and increased by about 2 ± 1.8 dB ($\mu \pm \sigma$, range: 0.7–4.1 dB) and 3.2 ± 0.8 dB ($\mu \pm \sigma$, range: 2.4–4.1 dB), respectively.

Interestingly, when using loudspeaker subset layouts with $N' \geq 8$ in the anechoic chamber, representing an ideal reproduction environment, the CS increases faster in the high frequency range with rather saturated values in the low-mid frequency range, cf. Table 4.5, which can be traced back to the concept of N'-CTC. With reference to Equation (3.5), the CTC matrix corresponds to an underdetermined system for $N' > 2$, optimised to exhibit minimal energy. A better spectral energy distribution in the CTC filters can be achieved by increasing N', which results in flatter filter curves that contain smaller shares of energy and less pronounced spectral peaks. The filters thus require less regularisation, allowing to better address the monaural cues in the playback HRTFs. This effect was mitigated but also present in the achievable in-situ CS. Returning to the perceptual demands regarding minimum CS (Parodi & Rubak, 2011), the average achievable in-situ CS obtained for the loudspeaker subset selections in the VR laboratory suggests to use the full loudspeaker layout in order to approach the recommended performance values. On the other hand, the subsets with $N' \geq 8$, investigated for the setup in the anechoic chamber, allow to largely meet the minimum perceptual CS thresholds. Once again, it should be noted that the perceptually exploitable CS may be elevated towards the optimal in-situ CS values.

Finally, it should be pointed out that the perceptual quality of CTC systems needs to be rated accounting for factors other than CS alone, such as spectral fidelity (Choueiri, 2008), phase issues, sensitivity to sweet spot offsets (Parodi & Rubak, 2010) and dynamic range control (Lentz, 2008; Sanches Masiero, 2012).

To answer research question Q4.5, increasing the number of loudspeaker has a positive effect on the achievable in-situ CS and favours meeting the perceptual requirements. The required loudspeaker number tends to be higher for setups installed in non-ideal reproduction environments compared to ideal environments due to the detrimental effect of reflections, and was estimated as $N'_{VR} = 16$ and $N'_{AC} = 8$ under rigorous assessment criteria.

4.4.2 Recreated in-situ receiver transfer functions

By reference to the initially posed research question Q4.6 and the hybrid static reproduction concept, cf. Section 3.4.2, the recreated receiver directivities need to analysed at the level of the ear canal entrances using the different spatial audio reproduction methods, cf. Section 3.3, implemented in the two experimental environments with 16-channel and 68-channel loudspeaker arrays, cf. Pausch (2022, Sec. 3.6) and Pausch, Behler, and Fels (2020).

Evaluation procedure

As reference condition, the resulting loudspeaker signals of each reproduction method, cf. Figure 3.4.2, were convolved with the ideal spatial playback transfer functions without room reflections, cf. corresponding paragraph in Section 4.2.2, of all involved loudspeakers N', and summed up per ear, representing the recreated ideal transfer functions. The loudspeaker signals either depend on the binaural input signal from a certain direction of incidence (CTC) or a panning direction $\boldsymbol{\theta}_p$ (VBAP, MDAP, HOA). In either case, the panning direction-dependent spatial rendering transfer functions without containing the free-field HA characteristics were selected to serve as target HRTFs and HARTFs, cf. Section 4.2.1. Note that CTC-based reproduction of simulated HA signals is not intended for the two specific system implementation variants presented in Section 3.4.1 and 3.4.2, but is only used for the current evaluation. For the panning-based methods, the resulting ear signals were energetically normalised by $1/\sqrt{N'}$. To get a first impression of the resulting ear signals, an example direction of incidence (panning direction $\boldsymbol{\theta}_p$) is plotted and discussed. In a global analysis, a grid of (panning) directions will be subsequently simulated and evaluated in terms of the resulting binaural cues. To see whether these cues can be adequately recreated at the centre of the physical array implementations, the previously used ideal spatial playback transfer functions are replaced by the short in-situ playback transfer functions, without containing the free-field HA microphone characteristics, to re-calculate the resulting binaural cues, representing the recreated in-situ transfer functions. In this way it is possible to compare the ideal and in-situ ear canal signals and infer influences of the respective reproduction environment.

System configuration

First, the specific parameter settings of each reproduction method are summarised. For N'-CTC reproduction, the 16-channel loudspeaker layouts, cf. Figure 4.12, were selected for both setups. All CTC filters were calculated on the basis of ideal playback HRTFs, and ideal front and rear HARTFs ($\beta = 0.01$), to be either applied on the ideal or short in-situ playback transfer functions. The generic rendering HRTFs and HARTFs from direction $\boldsymbol{\theta}_p$, representing the binaural input signals after convolution with a unit impulse, serve as binaural input signal to the CTC filter network, resulting in the loudspeaker signals. The loudspeaker signals for VBAP and MDAP were directly derived by applying the respective set of loudspeaker gains $\mathbf{g}$ on the loudspeaker signals (Politis, 2021b). A number of

20 spreading sources arranged on single concentric circles around $\boldsymbol{\theta}_{\mathrm{p}}$ with spreading angles of approximately 25.2 deg and 10.7 deg, see Figure 3.6b, determined the VBIP weights for MDAP (Politis, 2021b) applied to the 16-channel and 68-channel loudspeaker setups, respectively. The spreading angles were calculated as half the minimum loudspeaker span angles. Both MDAP spreading parameters were selected to achieve minimal variation of $\|\boldsymbol{r}_{\mathrm{E}}\|_2$ for varying panning directions (Frank, 2013; Zotter & Frank, 2019). For HOA-based reproduction, a spatial unit impulse at $\boldsymbol{\theta}_{\mathrm{p}}$ was encoded using SH truncation orders of $\mathcal{N}_{\mathrm{VR}} = 3$ and $\mathcal{N}_{\mathrm{AC}} = 7$, see Equation 3.15. Decoding relied on the AllRAD+ with max $\boldsymbol{r}_{\mathrm{E}}$ weighting, based on virtual loudspeaker layouts with t-design orders of t $= \mathcal{N} + 1$ and decoding orders matching the respective encoding orders (Politis, 2021a). Since the triangulations for directions below the lowest horizontal ring in the 16-channel loudspeaker setup exhibit loudspeaker span angles larger than 90 deg, the missing loudspeaker at the south pole was re-inserted as imaginary loudspeaker and included when calculating the triangulations and loudspeaker gains for VBAP (AllRAD+) and VBIP (MDAP). The imaginary loudspeaker weight was discarded before normalising the set of remaining gains. Assuming an average head radius of 8.75 cm (Hartley & Fry, 1921; Algazi, Avendano, & Duda, 2001), maximum frequency limits of about 1.9 kHz and 4.4 kHz, above which spatial aliasing occurs, can be estimated for the two setups using HOA (Bertet, Daniel, & Moreau, 2006).

Single-direction results

Figure 4.14 presents the magnitude spectra of recreated ideal receiver transfer functions for an example direction, applying the different spatial audio reproduction methods and the loudspeaker arrays in the VR laboratory and the anechoic chamber. The recreated ideal results using CTC coincide with the target results, since the CTC are perfectly matched in this evaluation scenario, except for the influence of regularisation. Spectral deviations owing to additional reflections will only become visible when applying the CTC filters on the short in-situ playback transfer functions. Overall, substantial deviations in magnitude spectra from the target magnitude spectra can be observed in all panning-based methods. In the first setup, cf. Figure 4.14a–4.14b, the ipsilateral HRTFs are more accurately reproduced by VBAP and MDAP, whereas HOA shows substantial spectral deviations already in the low frequency range, all methods exhibiting fair notch reproduction accuracy. However, all panning-based reproduction variants suffer from large notch mismatch errors on the contralateral ear signals. The description of these results can be largely transferred to the recreated ideal front HARTFs. In this scenario, the panning-based methods create erroneous notches at about

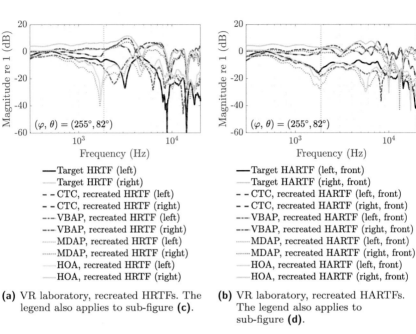

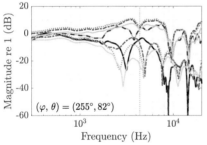

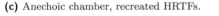

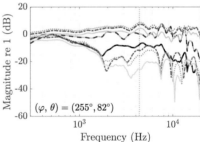

(a) VR laboratory, recreated HRTFs. The legend also applies to sub-figure **(c)**.

(b) VR laboratory, recreated HARTFs. The legend also applies to sub-figure **(d)**.

(c) Anechoic chamber, recreated HRTFs.

(d) Anechoic chamber, recreated front HARTFs.

Figure 4.14: Recreated ideal receiver transfer functions based on the corresponding ideal playback transfer functions for an example direction of incidence (panning direction θ_p) using the **(a)**–**(b)** 16-channel loudspeaker array in the VR laboratory, and the **(c)**–**(d)** 68-channel loudspeaker array in the anechoic chamber. The vertical dotted grey line represents the estimated spatial aliasing frequency.

10 kHz and 13 kHz, and overemphasise the spectral energy for frequencies larger than about 14 kHz. Slightly better results were achieved using the second setup, cf. Figure 4.14c–4.14d, with similarly high spectral deviations in the contralateral ear signals in panning-based methods but better recreated high-frequency characteristics in the front HARTFs. Interestingly, the HRTF notches around 9 kHz in the ipsilateral ear signals are partially well matched by the panning-based methods including HOA, even being above the spatial aliasing frequencies, indicated by the vertical grey dotted lines.

Global results

How the observed spectral deviations and additional reflections in the experimental setups affect binaural cues is now analysed by calculating the recreated receiver directivities for directions lying on a directional grid of 5 deg × 5 deg in azimuth and elevation angles, applying ideal and short in-situ playback transfer functions.

Spectral differences. Richter (2019) re-formulated the SD by Middlebrooks (1999) to obtain a frequency-dependent measure, defined as

$$\mathrm{SD}(f) = \sigma_w\left(20\log\left(\left|\frac{H_1(k,\vartheta,\varphi)}{H_{\mathrm{ref}}(k,\vartheta,\varphi)}\right|\right),w\right),\qquad(4.6)$$

representing the standard deviation of the logarithmic spectral error magnitude between an example receiver transfer function H_1 and the corresponding ideal transfer function H_{ref} in dB, spatially weighted by the Voronoi weights w (Keiner, Kunis, & Potts, 2007) and calculated per ear side. Figure 4.15 presents the left-ear SD, with columns and rows corresponding to the two experimental environments and the different spatial audio reproduction methods, respectively. The colors and line types indicate the types of receiver transfer functions, i.e. HRTFs, front and rear HARTFs, and the corresponding playback transfer functions, which the results are based on, i.e. ideal and short in-situ playback transfer functions, respectively.

For CTC-based reproduction, the SD in the recreated receiver transfer functions increases substantially when using short in-situ playback transfer functions for the CTC filter calculation in both experimental environments. Owing to the acoustically optimal nature in the anechoic chamber, the effect of reflections is not so strong but seems to particularly influence HRTFs.

Panning-based methods in both experimental environments exhibit an approximately constant SD bias in the low frequency range, hinting over-pronounced energy in the recreated spatial transfer functions. Across reproduction methods

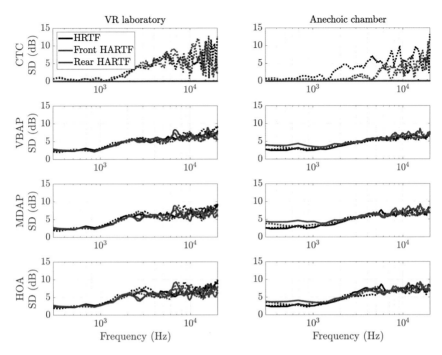

Figure 4.15: Left-ear HRTF / Left front and rear HARTF SDs in recreated receiver transfer functions based on the corresponding ideal and short in-situ playback transfer functions, indicated by solid and dashed lines, respectively, evaluated for directions of incidence (panning directions θ_p) lying on an equiangular grid with $5° \times 5°$ resolution in azimuth and elevation angles.

in the VR laboratory, a frequency limit of about 1 kHz can be identified as a common ground, above which a stronger increase in the SD can be observed. This error increase is less pronounced in the anechoic chamber with flatter slopes and less fluctuating curves towards higher frequencies. Moreover, similar curve progressions are further noticeable for the panning-based methods within each experimental environment, already entailing increased SD in the low frequency range below 1 kHz.

Binaural cues.   Being most relevant for the perception of VAEs, the recreated binaural cues and their deviations in ITDs and ILDs, symbolised by ΔITD and ΔILD, are evaluated as signed errors between the recreated data set and the reference data set, cf. Section 4.2.1. The ITDs were again estimated based on the IACC (Kistler & Wightman, 1992; Xie, 2013) in a frequency range between

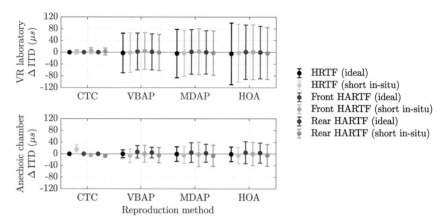

Figure 4.16: ITD errors in recreated HRTFs, and front and rear HARTFs, based on the corresponding ideal and short in-situ playback transfer functions, evaluated for directions of incidence (panning directions θ_p) lying on an equiangular grid with $5° \times 5°$ resolution in azimuth and elevation angles. Bars and error bars represent means and standard deviations, respectively.

300 Hz and 1.5 kHz. ILD deviations, cf. Equation (2.16), were evaluated for frequencies $1.5\,\text{kHz} \leq f \leq 20\,\text{kHz}$ by means of spectral division and subsequent subtraction of the magnitude values obtained for the reference dataset. Statistical hypothesis tests were deliberately omitted, as the very high statistical power entails that even small differences in mean values would lead to significant but perceptually negligible results. Instead, the key tendencies and variance differences in the principal factors of interest, as per the initially posed research question Q4.6, will be analysed briefly for both binaural performance measures across experimental factors below.

Figure 4.16 shows average errors with standard deviations in ΔITD after removing outliers above and below 1.5 times the upper and lower quartile ranges, respectively. The results are split into experimental environments (VR laboratory, anechoic chamber), receiver transfer function type (HRTF, front HARTF, and rear HARTF), playback transfer function type (ideal, and short in-situ), and reproduction type (CTC, VBAP, MDAP, and HOA). A strongly increasing error variance in panning-based methods compared to CTC reproduction shall be noted, especially in the VR laboratory. Slight gradual variance increases can be seen from VBAP, over MDAP to HOA. Reasons for a more moderate error increase using panning-based methods in the anechoic chamber can again be partly attributed to the room type, but particularly to the higher number of loudspeakers and

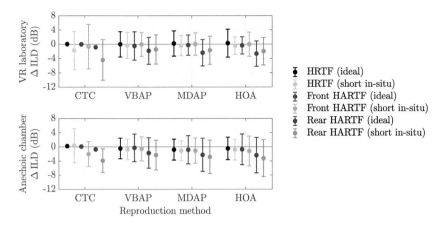

Figure 4.17: ILD errors in recreated HRTFs, and front and rear HARTFs, based on the corresponding ideal and short in-situ playback transfer functions, evaluated for directions of incidence (panning directions $\boldsymbol{\theta}_\mathrm{p}$) lying on an equiangular grid with $5° \times 5°$ resolution in azimuth and elevation angles. Bars and error bars represent means and standard deviations, respectively.

the associated HOA decoder order, both allowing refined panning and increased directional performance regarding ITD recreation.

No obvious changes in means and variances of ΔITD are visible in the VR laboratory across reproduction methods when using short in-situ playback transfer functions. In comparison, the ITD errors seem to be slightly more susceptible to mismatched playback transfer functions in the anechoic chamber. The strongest effect can be observed on HRTFs in CTC-based binaural reproduction, while panning-based methods still entail comparable error variations and means fluctuating around zero.

Average ILD errors with standard deviations, with outliers removed as done for ΔITD, are shown by experimental environments in Figure 4.17, again split into the levels of experimental factors. Compared to ΔITD, the error patterns are not that consistent. The means and variances of this error metric are much more affected in CTC-based reproduction when using short in-situ playback transfer functions, distorted by detrimental reflections, what was already foreseeable in the light of the SD results, cf. Figure 4.15. Notably higher ILD variances and changes in means can be observed in rear HARTFs. Reasons for this error increase may be due to remaining differences in the positioning of the research HAs across measurement sessions, although special care was taken to preserve the marked position. In general, a mismatch in playback transfer function will also likely impact the CS and localisation performance (Majdak, Masiero, & Fels, 2013).

Roughly similar variances and tendencies are present in panning-based methods in both experimental environments, with a tendency of ILD underestimation in the anechoic chamber.

Collectively, the results regarding ΔITD and ΔILD provide an estimation of the achievable binaural cue preservation performance across reproduction systems using ideal and in-situ playback transfer functions. Special attention must be paid to the calibration of the real-world head pose of the listener in CTC-based binaural reproduction, otherwise substantial errors due to mismatched playback transfer functions and corresponding CTC filters can be expected, continuing to deteriorate when already using generic datasets for CTC filter generation. Applying laser measurement tools, head rests in static listening experiments, and motion tracking systems in interactive auralisation systems can be considered good practice to minimise the mismatch error.

To provide an answer to research question Q4.6, it can be summarised that due to the underlying principles, panning-based methods will generally entail higher errors in recreated binaural cues. This tendency was particularly pronounced in ΔITD in non-ideal reproduction environments with a lower number of loudspeakers and poorer available directional mapping flexibility.

5

Perceptual system evaluation

The objective evaluations of the auralisation systems, described in Chapter 4, are ideally complemented by perceptual experiments. Specific implementations and the underlying concepts of spatial audio reproduction systems usually make it necessary to conduct specially designed studies for comparative analyses, even though there also exist well-developed auditory models (see, e.g., Søndergaard & Majdak, 2013; Georganti et al., 2013; Baumgartner, Majdak, & Laback, 2014) for the prediction of spatial audio quality parameters (Lindau et al., 2014; Nicol et al., 2014; Raake, Wierstorf, & Blauert, 2014; Simon, Zacharov, & Katz, 2016). While comprehensive system evaluations are usually not feasible, partial assessment already allows the determination of previously unknown and potentially influential shortcomings. If the auralisation system is capable of generating VSSs at arbitrary positions in the virtual space it is particularly interesting how accurately these sources can be perceptually resolved within the egocentric spatial domain, which is investigated in the two experiments presented in this chapter. Experiment 1 addresses the angular component and examines the localisation of RSSs in comparison to VSSs, reproduced using interactive binaural reproduction involving research HAs, in free-field conditions. System-inherent perceptual properties are identified and discussed on the basis of reversal rates, overall horizontal localisation performance and angular error measures. The focus of Experiment 2 is on the radial component, in which participants were asked to rate the perceived egocentric distance of VSSs in simulated room acoustics. It is investigated how well this spatial audio quality parameter can be conveyed by means of VSSs that are either binaurally reproduced via headphones, research HAs, loudspeak-

Related publications:

Pausch et al., 2018b; Pausch and Fels, 2020.

ers using CTC filters, or via panning-based methods. Combined playback with research HAs was additionally tested with the aforementioned spatial audio reproduction methods, except with headphones. The distance estimates are fitted by use of a compressive power function model and analysed in terms of the calculated auditory horizons and root mean square (RMS) errors. The results from both studies should provide baselines on how exactly the respective spatial audio quality parameters can be conveyed by the considered reproduction methods.

5.1 Experiment 1: Localisation performance in a binaural real-time auralisation system extended to research hearing aids

Localisation of spatialised sound sources represents one of the most important parameters for navigation in real-life environments. For this reason it is particularly important that auralisation systems, aiming to replicate such real-life environments by means of VAEs, are capable of granting access to the two most important binaural cues for source localisation in the horizontal plane, that is, ILDs and ITDs (Rayleigh, 1907; Møller, 1992; Blauert, 1997), see also Sections 2.8.1 and 2.8.2. If sources are elevated or positioned on cones of confusion, additional monaural cues help to determine their position and help to reduce erroneous localisation in the opposite vertical hemisphere (Musicant & Butler, 1985; Wightman & Kistler, 1997). Similar improvements in localisation accuracy can be achieved by exploiting dynamic binaural cues, which are accessed through head movements (Thurlow & Runge, 1967; McAnally & Martin, 2014). Experiment 1 aimed to clarify how well the binaural real-time reproduction system, see Section 3.4.1, is capable of replicating these localisation cues.

The current study

The different experimental conditions of Experiment 1 are concisely outlined below. By reference to the experiments of Bronkhorst (1995), the localisation performance of VSSs is assessed and compared to RSSs, modelled by discrete loudspeakers. Representing a potential test and training environments for studies involving HA users (Cameron & Dillon, 2011; Cameron, Glyde, & Dillon, 2012), it is of particular interest how well VSSs can be localised when reproduced via open-fit research HAs. An objective evaluation of monaural cue preservation was carried out by Denk, Ewert, and Kollmeier (2018) for different HA device styles on the basis of auditory models predicting the localisation performance in the saggital plane (Baumgartner, Majdak, & Laback, 2014). Inspired by the experiment of Mueller et al. (2012), who tested localisation of VSSs based on generic HARTFs in adults fitted with open-fit BTE HAs while immersed in complex scenes with room acoustics, the current study tested free-field localisation using similar devices, operated in omnidirectional mode. Majdak, Masiero, and Fels (2013) investigated the performance difference in matched and mismatched virtual CTC system configurations. For the current study, binaural reproduction via loudspeakers with CTC filters was extended by including research HAs. Two main research questions shall be answered:

Q5.1 In comparison to the localisation of RSSs, reproduced via discrete loud-speakers, how well can VSSs be dynamically localised when reproduced over headphones, loudspeakers with CTC filters, research HAs alone, or combined via loudspeakers with CTC filters and research HAs?

Q5.2 Does combined binaural playback via loudspeakers with CTC filters and research HAs improve the localisation performance compared to playback via research HAs alone?

Since the system was designed for HA research, the perceptual results will set a perceptual benchmark for upcoming evaluations and applications. With this in mind, the system can be considered a helpful tool for the development of standard-ised procedures to evaluate HA algorithms for devices with open fitting, while allowing to drive the development of advanced fitting routines in dynamically reproduced VAEs.

5.1.1 Methods

Participants

The experiment was conducted involving fifteen non-expert adults (9 female) aged 24 ± 5.4 ($\mu \pm \sigma$, range: 18–35) with self-reported NH, no history of HL, and normal (or corrected-to-normal) vision. All participants were reimbursed for their participation and provided written informed consent prior to the experiment. European and country-specific data protection regulations were complied with in terms of handling and storing data (European Union, 2016).

Stimulus

A single-channel white noise train with unwindowed on- and offsets, consisting of two noise pulses with a duration of $1\,\mathrm{s}$ each and an intermediate pause of $0.25\,\mathrm{s}$ was generated in MATLAB™ at $f_\mathrm{s} = 44.1\,\mathrm{kHz}$. Such a stimulus length allows the application of head movements during playback for dynamic binaural cue access, enabling the highest possible localisation accuracy (Thurlow & Mergener, 1970).

Virtual sound sources

For the creation of VSSs, the generic spatial rendering HRTF and HARTF datasets, cf. Section 4.2.1, were used for this study. The applied filter sets had a length of 256 samples at $f_s = 44.1\,\mathrm{kHz}$ with a spatial resolution of $1\,\mathrm{deg} \times 1\,\mathrm{deg}$ in azimuth and elevation angles, fulfilling the ideal requirement of lying in the range or below the minimum audible angle (Mills, 1958; Perrott & Pacheco, 1989). I used the real-time auralisation software framework VIRTUAL ACOUSTICS (2021) to create the VSSs by convolving the stimulus with the respective directional dataset, that is HRTFs or HARTFs, as required per experimental condition. The directional transfer functions were selected via a nearest-neighbour algorithm, depending on the current real-world listener pose relative to the VSS, as captured by the optical motion tracking system, see Pausch (2022, Sec. 3.1). For efficient time-varying filtering, a time-domain cross fading technique was applied to exchange the filters in real time (Wefers, 2015). In case of changing the current pose by $\pm 0.5\,\mathrm{deg}$ relative to the VSS, the auralisation system provided maximum filter update rates of about $172\,\mathrm{Hz}$ at an audio buffer size of 256 samples and $f_s = 44.1\,\mathrm{kHz}$.

Experimental design and test conditions

Designed as a closed-loop experiment[1], the participants had to rate the perceived directions of twelve sound sources, eight of which were arranged at $\varphi_i = i \cdot 45°$, with $i \in \{0, 1, \ldots, 7\}$, in the horizontal plane, and four sound source directions in the median plane, two each at φ_0 and φ_4 and elevation angles of $\vartheta = \{30°, -30°\}$. Each source direction was tested three times with random presentation order. The experiment was performed on two different days and corresponded to a within-participant design, consisting of two parts which included two and three test blocks, respectively. The test blocks were counterbalanced following a Latin square design to avoid first-order carryover effects (Williams, 1949). Both parts of the experiment aimed to explore the effects of the within-participant factor *System*, consisting of the factor levels as described below. The first part was performed by Richter (2019) in the anechoic chamber, see Pausch (2022, Sec. 2.5), investigating the localisation accuracy of RSSs (level LS) and VSSs over headphones (level HP). The second part took place in a commercial hearing booth, see Pausch (2022, Sec. 2.1), measuring the localisation performance of VSSs binaurally reproduced via loudspeakers with CTC filters (level CTC), research HAs (level RHA), and via loudspeakers with CTC filters in combination with research HAs (level CTCwRHA). Table 5.1 summarises these experimental conditions.

---

[1]Incorrectly classified as *open-loop* experiment in Pausch and Fels (2020).

Table 5.1: Experiment 1: Test conditions, specified as levels of the within-participant factor *System*, and listening environments with playback devices and sound source types.

Index	Reproduction environment	Condition	Playback device (source type)
1	Anechoic chamber	LS	Discrete loudspeakers (real)
2	Anechoic chamber	HP	Headphones (virtual)
3	Commercial hearing booth	CTC	Loudspeakers with CTC filters (virtual)
4	Commercial hearing booth	RHA	Research HAs (virtual)
5	Commercial hearing booth	CTCwRHA	Loudspeakers with CTC filters and research HAs (virtual)

Apparatus

The first part of this experiment was conducted by Richter (2019) using a previous loudspeaker setup installed in the anechoic chamber (Oberem et al., 2014). In condition LS, the RSSs were modelled by playing back the stimuli directly via one of twelve two-way loudspeakers (Genelec 6010, Genelec Oy, Iisalmi, Finland). With the aim to reduce the influence of the headphone transducer characteristics (HD 600, Sennheiser, Wedemark, Germany) in condition HP, a perceptually robust headphone equalisation method (Masiero & Fels, 2011), described in Pausch (2022, Sec. 3.3), was utilised to equalise the binaural signals before playback. Only the data from the subset of participants who also took part in the second part of the experiment, with conditions as specified below, were included in the data analysis.

Signal playback in conditions CTC relied on the four loudspeakers installed in the commercial hearing booth, with azimuth angles as described in Pausch (2022, Sec. 3.5), but placed at a common elevation angle of $\vartheta = 20°$ (Parodi & Rubak, 2010), with the individual loudspeaker orientations set in a way that their main axes intersected with the centre of the listener's ear axis. For robust binaural reproduction, a 4-CTC approach was used with the CTC system matrix regularised using a Tikhonov regularisation factor of 0.01 (Sanches Masiero, 2012). In condition RHA, the HARTFs of the front microphones of the research HAs, see Pausch (2022, Sec. 3.4), were used for the generation of VSSs, to be played back via the HA receivers. For combined reproduction in condition CTCwRHA, a time delay was configured, so that the binaural signals of the research HAs arrive 7 ms after the loudspeaker-based binaural signals (Stone et al., 2008). To verify this relative delay, I conducted IR measurements using the custom-made artificial head, see Pausch (2022, Sec. 3.2), to estimate the TOAs, as done in

Section 4.3.2. This measurement was performed separately for the conditions CTC and RHA, each time accounting for the rendering and reproduction delays.

Any listener pose changes were captured by the electromagnetic and the optical tracking systems, see Pausch (2022, Sec. 3.1), in the first and second part of the experiment, respectively, set to a maximum common frame rate of 60 Hz. The offset of the rigid body mounted on top of the participant's head to the centre of the interaural axis was corrected individually (Berzborn et al., 2017). Accounting for the measured mean calculation times for auralising the DS in the binaural real-time system implementation, see Section 4.3.1, the dynamic end-to-end latency can be expected to be well below the minimum detectable threshold values (Brungart, Simpson, & Kordik, 2005; Yairi, Iwaya, & Suzuki, 2006; Lindau, 2009).

To avoid a perceptual bias, a common reproduction level of 65 dB(A) was set for all experimental conditions, based on calibrated measurements obtained from the same custom-made artificial head. While individual gains were applied in conditions CTC and RHA, condition CTCwRHA required an additional combined gain of -3 dB to match the target reproduction level. For an analysis of the individual contribution by each reproduction path in CTCwRHA, the SPLs were measured via the artificial head with ear simulator (HMS III, HEAD acoustics, Herzogenrath, Germany), see Pausch (2022, Sec. 3.2), used in combination with a conditioning amplifier (Type 2690-A, Nexus, Brüel & Kjær, Nærum, Denmark) and an audio interface (RME Fireface UC, Audio AG, Haimhausen, Germany), in two in-situ measurement scenarios. The measurements were performed at 25.4 °C air temperature and a relative humidity of 33 %. In both scenarios, the artificial head was placed in the centre of the hearing booth at the nominal ear height of 1.2 m to sequentially measure the resulting reproduction level in condition CTC, produced by the 12 VSSs playing back the stimulus from the corresponding direction. The same measurement was subsequently performed for condition HA. The measured SPLs in third-octave bands with centre frequencies between 62.5 Hz and 16 kHz, averaged across all VSS directions, are shown in Figure 5.1. While the spectral level decay in condition CTC roots in the properties of HRTFs, the distinct peaks at 155 Hz and 250 Hz can be attributed to the influence of room modes in the reproduction environment. Note that the passive damping properties of the open silicone dome attached to the HAs affect the SPLs in the third-octave bands with centre frequencies above 2 kHz, cf. Pausch (2022, Fig. 8). Prominent energy accumulations at the first and second ear canal simulator resonance frequencies around 3 kHz and 9 kHz can be identified in condition RHA, in addition to the band limitation of the HA receiver when used in combination with an open-dome ear piece, see Pausch (2022, Fig. 9b).

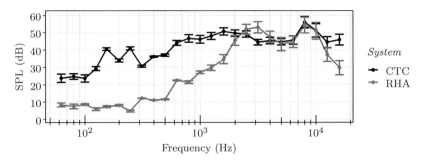

Figure 5.1: Measured SPLs in third-octave bands, as contributed by the individual reproduction paths in condition CTCwRHA at artificial ear drum level, averaged
across VSS directions. Error bars represent the 95 % confidence intervals
(CIs) of the means. (Figure taken from Pausch and Fels, 2020, licensed under
CC BY 4.0, and adapted.)

Pointing method

An exocentric pointing method (Richter & Fels, 2016) was applied to rate the
perceived sound source directions. A monitor in front of the participants displayed a GUI, showing a sphere with the horizontal and frontal plane indicated
by great circles (orthodromes) and the participant's nominal virtual viewing directions represented by a three-dimensional arrow in the centre of this sphere.
Twenty vertical and 20 horizontal 1 deg × 1 deg squares constructed a crosshair
with the centre square indicating the perceived sound source direction. The
three-dimensional representation of this crosshair was improved by a surrounding and pursuant grid, which spanned a spherical lune (biangle) in the region
of the currently indicated direction, consisting of squares of 5 deg × 5 deg each.
The participants used a game controller (F710, Logitech, Romanel-sur-Morges,
Switzerland) to move a crosshair horizontally and vertically via operating the
left and right joystick, respectively, and pressed the green button to log in their
responses. For sound source directions in the rear hemisphere, i.e. in the angular
range between $90° \leq \varphi \leq 270°$, it was possible to invert the virtual viewing
direction using the blue button. Comparing the localisation errors using this
pointing method with the ones resulting when using nose pointing, Richter and
Fels (2016) did not report significant differences.

Experimental procedure

Written informed consent was obtained from all participants before each part of
the experiment, after they had been informed about the experimental procedure
and other relevant information, including their rights. Thereafter, the individual

headphone transfer functions were measured. In the training sessions, conducted before each test session as per the current experimental condition, see Table 5.1, the participants were familiarised with the procedure and pointing method by randomly testing ten sound source directions. To promote learning, the presented sound source direction was marked by a red 10 deg × 10 deg square within which area participants had to navigate the crosshair centre and confirm the perceived direction. In consecutive training trials, the difficulty had been gradually decreased by reducing the target area to a 1 deg × 1 deg square. After having completed the training, all source directions were tested three times in a randomised way, without showing the target direction in the GUI. It was allowed to repeat the presented sound source directions twice per trial using the red button of the gamepad. In total, a number of 12 directions × 3 repetitions × 5 blocks = 180 trials had to be rated by each participants. After each condition, a forced break of 5 min was imposed. On average, the participants needed about 45 min and 60 min to complete the first and second part of the experiment, respectively.

Error metrics

Reversal rates. The correction of reversal rates accounted for a possible shift of the frontal plane due to head movements relative to the presented static sound source direction. Following a similar correction strategy as in Chen (2003), reversals were only corrected if the perceived direction lay ±30 deg around the presented direction, when mirrored along the frontal plane. In case of a reversal, this trial had been corrected by mirroring the perceived sound source direction on the frontal plane. No correction was applied for directions at $\varphi = 90°$ and $\varphi = 270°$. The reversal rates were calculated as percentages for the ten relevant source positions by comparing the corrected with the uncorrected results, averaged across participants per experimental condition and split into front-back, back-front and pooled reversal rates.

Angular errors. The localisation errors were defined as the difference between the presented and perceived sound source directions, i.e. (φ, ϑ) and $(\hat{\varphi}, \hat{\vartheta})$, respectively, represented by three angular error metrics: azimuth error ϵ_φ, elevation error ϵ_ϑ and overall error

$$\epsilon_\gamma = \arccos\big(\cos(\vartheta)\cos(\hat{\vartheta})\cos(\varphi - \hat{\varphi}) + \sin(\varphi)\sin(\hat{\varphi})\big), \qquad (5.1)$$

the latter denoting the great circle central angle between presented and perceived sound source directions (G. T. M., 1932). The azimuth and elevation errors are orthogonal to each other. All angular error metrics are reported as unsigned errors below, representing error magnitudes in deg.

Table 5.2: Hypotheses of Experiment 1.

Code	Hypothesis
H1-Exp1	The localisation performance in conditions HP, CTC, RHA and CTCwRHA decreases in comparison with the performance observed in condition LS.
H2-Exp1	The localisation performance improves in condition CTCwRHA in comparison with the performance observed in condition RHA.

Hypotheses and data analysis

Based on the initial research questions, two main hypotheses were formulated, see Table 5.2. All data analysis and plotting of the results in Experiment 1 (and Experiment 2) were performed using R (R Core Team, 2021) and RStudio (RStudio Team, 2021). In both experiments, the results regarding statistical hypothesis testing are interpreted at a confidence level of 95 %. Reported bootstrapped results are based on 10,000 bootstrap samples.

5.1.2 Results

Reversal rates

Table 5.3 and Figure 5.2 present the percentages of front-back, back-front, and pooled reversal rates per experimental condition.

The lowest front-back confusions of 5.1 % and 4.4 % were observed for the conditions LS and CTC, respectively. The rate substantially increased to 13.6 % and the largest percentage of 17.6 % in the conditions HP and RHA, respectively, with a lowered rate of 10.7 % in condition CTCwRHA.

Compared to the front-back confusions, the back-front confusion values were lower in conditions LS, HP, RHA, and CTCwRHA with average values of 3.3 %, 2.9 %, 10.2 % and 8.2 %, respectively. A roughly inverted pattern between the conditions HP and CTC was observed, with average reversal percentages of 11.8 % in the latter condition.

Concerning pooled reversals, the lowest rates of 8.4 % were found for condition LS, increasing to 16.4 % and 16.2 % in conditions HP and CTC, respectively. Rating the VSS positions in condition RHA amounted to the highest reversal rate of 27.8 %, which substantially decreased to 18.9 % in condition CTCwRHA.

Overall horizontal source localisation

The overall horizontal source localisation was analysed based on the perceived directions of all sound sources in the horizontal plane, consolidated by the factor *Presented azimuth* with azimuth angles φ_i as factor levels. The perceived azimuth

Table 5.3: Summary of mean front-back, back-front and pooled reversal rates with standard errors (SEs), averaged across all participant trials.

	Reversal rate		
Condition	Front-back (%) $\mu \pm \sigma_{se}$	Back-front (%) $\mu \pm \sigma_{se}$	Pooled (%) $\mu \pm \sigma_{se}$
LS	5.1 ± 1.3	3.3 ± 1.8	8.4 ± 2.4
HP	13.6 ± 2.8	2.9 ± 1.3	16.4 ± 3.1
CTC	4.4 ± 1.9	11.8 ± 4.3	16.2 ± 4.7
RHA	17.6 ± 3.8	10.2 ± 3.1	27.8 ± 4.6
CTCwRHA	10.7 ± 3.7	8.2 ± 2.8	18.9 ± 4.2

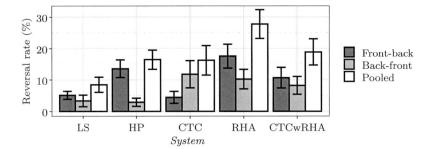

Figure 5.2: Mean front-back, back-front and pooled reversals rates per experimental condition. The error bars show one SE of the mean.

angles $\hat{\varphi}_i$, collectively referred to by the outcome variable *Perceived azimuth*, are shown as scatter plots in relation to the presented azimuth angles in Figure 5.3 per experimental condition, including fitted linear regression lines. Each grey dot represents the averaged by-direction results per participant with corrected reversals. While the dashed black lines indicate perfect agreement, the grey linear regression lines show the least square fits with bootstrapped 95 % confidence regions. Black dots with error bars further display the means and bootstrapped 95 % CIs of the means per azimuth direction. The linear regression lines are complemented by the regression equations and R^2 values.

To compare the overall horizontal source localisation performance within the scope of statistical hypothesis testing, the data analysis described below aimed at detecting differences in the intercepts and slopes of the regression lines between the experimental conditions. A linear mixed-effect (LME) model was formulated and fit by restricted maximum likelihood estimation (Bates et al., 2015) with unconstrained and bounds-constrained quasi-Newton method optimiser (Nash & Varadhan, 2011; Nash, 2014) to predict the outcome variable *Perceived azimuth*.

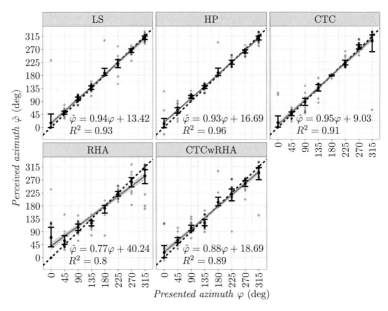

Figure 5.3: Localisation results with corrected reversals for source directions in the horizontal plane. The grey regression lines represent the least-square fits with bootstrapped 95 % confidence regions. Means and bootstrapped 95 % CIs per source direction are represented by black dots and error bars, respectively, showing averaged results over by-direction trials across participants. (Figure taken from Pausch and Fels, 2020, licensed under CC BY 4.0, and adapted.)

As fixed-effect terms, I entered the factors *Presented azimuth*, recoded as numeric factor, and *System* with its levels corresponding to the experimental conditions, including the interaction term. Random effects were addressed by entering the individual participant responses, averaged over the by-direction trials, as random by-participant intercepts. The nesting of participants within each level of *Presented azimuth* and *System* was modelled by two additional random-effect terms. Two models with identical structures but different reference levels in *System* were created to test the hypotheses, see Tables 5.2. In Model 1, condition LS was chosen as reference level, allowing to compare the intercept and slope differences with the remaining conditions, see H1-Exp1. Instead, Model 2 facilitates investigating whether the combined reproduction in condition CTCwRHA led to intercept and slope differences and helped to improve the overall horizontal source localisation in comparison to condition RHA, see H2-Exp1.

To obtain a parsimonious model, a backward elimination procedure with the aim of minimising the Akaike information criterion (AIC) was first applied on random-effect terms and consecutively on fixed-effect terms (Kuznetsova, Brock-hoff, & Christensen, 2017). The results of consecutive likelihood ratio tests on the random-effect structure classified the nesting of participants within *System*, $\chi^2(1) = 0.067$, $p = .795$ and the random by-participant intercept term, $\chi^2(1) < 0.001$, $p = .999$, as redundant, while suggesting to keep the nesting of participants within *Presented azimuth*, $\chi^2(1) = 37.27$, $p < .001$. To investigate whether dropping fixed effects showed improvements in the explained variance, *F*-tests using Kenward–Roger's method (Halekoh & Højsgaard, 2014) were performed, suggesting to preserve the interaction term *System* × *Presented azimuth*, $F(4, 472) = 8.89$, $p < .001$. The resulting final model had been improved from an AIC of 5807.92 to 5803.98 and is summarised in Table A.1 (Lüdecke, 2021). Normality assumptions were visually checked by comparing standardised residuals versus fitted values, which did not reveal any obvious violations. The marginal R^2 value indicated that the model with fixed effects alone explained 90 % of the variance in *Perceived azimuth*, while accounting for both fixed and random effects significantly improved the model performance, reflected by a conditional R^2 value of 92 %.

Planned comparisons in Model 1 resulted in an intercept difference in RHA versus LS, $t(472) = 4.13$, $p < .001$, and a slope effect of *Presented azimuth*, $t(505.83) = 34.24$, $p < .001$. However, the interpretation has to account for the significant interaction RHA versus LS × *Presented azimuth*, $t(472) = -4.82$, $p < .001$, which indicates that the estimated regression line slope of 0.77 in condition RHA is lower than the estimated value of 0.94 in condition LS.[2] This result partially supports hypothesis H1-Exp1 regarding the overall horizontal source localisation. The planned comparisons results revealed no other significant effects or interactions.

The results of planned comparisons applied to Model 2 indicated an intercept effect of RHA, $t(505.83) = 7.81$, $p < .001$, a slope effect of *Presented azimuth*, $t(505.83) = 28.15$, $p < .001$, and intercept differences in LS versus RHA, $t(472) = -4.13$, $p < .001$, in HP versus RHA, $t(472) = -3.62$, $p < .001$, in CTC versus RHA, $t(472) = -4.8$, $p < .001$, and in CTCwRHA versus RHA, $t(472) = -3.32$, $p < .001$. Any interpretations need to consider the significant interactions between LS versus RHA × *Presented azimuth*, $t(472) = 4.82$, $p < .001$, HP versus RHA × *Presented azimuth*, $t(472) = 4.51$, $p < .001$, CTC versus RHA × *Presented azimuth*, $t(472) = 5.15$, $p < .001$, and CTCwRHA versus RHA × *Presented azimuth*, $t(472) = 3.34$, $p < .001$. Relevant for hypothesis testing, the interaction CTCwRHA versus RHA × *Presented azimuth* indicated a

---

[2] Pausch and Fels (2020) erroneously related the result to condition HP, which, however, has no effect on the drawn conclusions in connection with statistical hypothesis testing.

Table 5.4: Summary of mean angular error metrics, calculated over averaged by-direction trials across participants.

	Angular error metric		
Condition	ϵ_φ (deg) $\mu \pm \sigma_{\mathrm{se}}$	ϵ_ϑ (deg) $\mu \pm \sigma_{\mathrm{se}}$	ϵ_γ (deg) $\mu \pm \sigma_{\mathrm{se}}$
LS	13.2 ± 6.2	11.3 ± 1.8	16.6 ± 2.3
HP	$17 \ \pm 3.9$	14.6 ± 1.4	23.6 ± 2.4
CTC	$21 \ \pm 4.5$	21.7 ± 1.6	29.1 ± 1.5
RHA	29.6 ± 5.5	21.9 ± 2.0	39.3 ± 3.3
CTCwRHA	20.3 ± 3.6	22.7 ± 2.2	33.8 ± 2.0

higher regression line slope of 0.88 in condition CTCwRHA compared to a slope of 0.77 in condition RHA and therefore supports H2-Exp1 in terms of overall horizontal source localisation.

Angular errors

Table 5.4 and Figure 5.4 present the results on angular error metrics, calculated based on the perceptual ratings of all 12 sound source directions, averaged over by-direction trials across participants per condition. Per panel, the average and median results per experimental condition are plotted as black dots and grey crosses, respectively, together with error bars, representing the bootstrapped 95 % CIs of the mean.

Given that the data complexity has been reduced to clusters aggregating the angular error results per the levels of *System*, I switched to a simpler data analysis strategy by means of ANOVA. To check for fulfillment of the assumptions, Shapiro–Wilk tests were conducted and indicated that 90 % of the log-transformed data exhibited normally distributed residuals. Since there is evidence that ANOVA is sufficiently robust against violations of the normality assumption (Pearson, 1931; Schmider et al., 2010), three one-way repeated-measures ANOVAs were conducted to assess the effect of the within-participant factor *System* on each angular error metric. Multiple t-tests with Holm–Bonferroni correction (Holm, 1979) were performed as required for statistical hypothesis testing. These planned comparisons are represented by the letter Δ with indices referring to the corresponding experimental condition (e.g., Δ_{31} indicating the comparison between conditions CTC and LS), see first column of Table 5.1.

A one-way repeated-measures ANOVA with type III sum of squares indicated a significant effect of *System* on ϵ_φ, $F(4, 56) = 7.88$, $p < .001$, $\eta_{\mathrm{p}}^2 = .36$. Planned comparisons revealed a significant increase of ϵ_φ in Δ_{31}, $t(56) = -3.83$, $p = .002$ (CTC, $\mu = 20.95$, $\sigma_{\mathrm{se}} = 4.54$; LS, $\mu = 13.24$, $\sigma_{\mathrm{se}} = 6.19$), Δ_{41}, $t(56) = -5.29$,

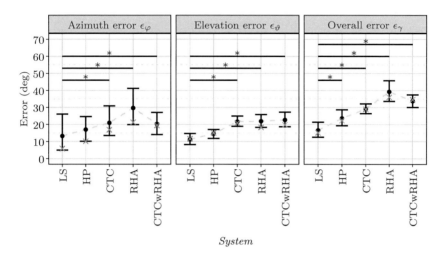

Figure 5.4: Unsigned localisation errors ϵ_φ, ϵ_ϑ and ϵ_γ, averaged over by-direction trials across participants per condition in Experiment 1. The means with bootstrapped 95 % CIs are represented by black dots and error bars, respectively, and medians are indicated by crosses. The horizontal lines with asterisks indicate significant mean differences at the 95 % confidence level.

$p < .001$ (RHA, $\mu = 29.64$, $\sigma_{se} = 5.51$), and Δ_{51}, $t(56) = -3.72$, $p = .002$ (CTCwRHA, $\mu = 20.28$, $\sigma_{se} = 3.58$). These results partially support H1-Exp1 regarding ϵ_φ, see Table 5.2. No support of H2-Exp1 in terms of ϵ_φ was found due to the absence of other significant differences.

A one-way repeated-measures ANOVA with type III sum of squares indicated a significant effect of *System* on ϵ_ϑ, $F(1.52, 21.33) = 8.44$, $p = .004$, $\eta_p^2 = .38$. Since the results of Mauchly's test pointed out a violation of the sphericity assumption had been violated, $\chi^2(9) = 45.60$, $p < .001$, Greenhouse-Geisser estimates of sphericity, $\varepsilon = .38$, were applied to correct the degrees of freedom. The planned comparisons revealed a significant increase of ϵ_ϑ in Δ_{31}, $t(56) = -4.62$, $p < .001$ (CTC, $\mu = 21.66$, $\sigma_{se} = 1.59$; LS, $\mu = 11.28$, $\sigma_{se} = 1.76$), Δ_{41}, $t(56) = -4.56$, $p < .001$ (RHA, $\mu = 21.93$, $\sigma_{se} = 2.04$), and Δ_{51}, $t(56) = -4.72$, $p < .001$ (CTCwRHA, $\mu = 22.66$, $\sigma_{se} = 2.16$). These results partially support H1-Exp1 regarding ϵ_ϑ. No support of H2-Exp1 in terms of ϵ_ϑ was found due to the absence of other significant differences.

A one-way repeated-measures ANOVA with type III sum of squares revealed a significant effect of *System* on ϵ_γ, $F(2.29, 32.03) = 21.23$, $p < .001$, $\eta_p^2 = .60$. Mauchly's test results indicated a violation of the sphericity assumption, $\chi^2(9) = 19.30$, $p = .025$, Greenhouse-Geisser estimates of sphericity, $\varepsilon = .57$, were

Table 5.5: Experiment 1: Results of planned comparisons on angular error metrics. Asterisks denote significant differences at the confidence level of 95 %.

Contrast	Angular error metric		
	ϵ_φ	ϵ_ϑ	ϵ_γ
Δ_{21} (HP vs. LS)	n.s.	n.s.	*
Δ_{31} (CTC vs. LS)	*	*	*
Δ_{41} (RHA vs. LS)	*	*	*
Δ_{51} (CTCwRHA vs. LS)	*	*	*
Δ_{54} (CTCwRHA vs. RHA)	n.s.	n.s.	n.s.

applied to correct the degrees of freedom. The results from planned comparisons suggested a significant increase of ϵ_γ in Δ_{21}, $t(56) = -3.52$, $p = .004$ (HP, $\mu = 23.61$, $\sigma_{se} = 2.39$; LS, $\mu = 16.45$, $\sigma_{se} = 2.32$), Δ_{31}, $t(56) = -5.83$, $p < .001$ (CTC, $\mu = 29.09$, $\sigma_{se} = 1.53$), Δ_{41}, $t(56) = -8.2$, $p < .001$ (RHA, $\mu = 39.26$, $\sigma_{se} = 3.25$), and Δ_{51}, $t(56) = -7.08$, $p < .001$ (CTCwRHA, $\mu = 33.76$, $\sigma_{se} = 1.95$). These results partially support H1-Exp1 regarding ϵ_γ. Since no other significant differences were found, H2-Exp1 could not be substantiated in terms of ϵ_γ.

Table 5.5 summarises the results of the planned comparisons.

5.1.3 Discussion

Reversal rates

In wide accordance to the results of previous studies, the lowest pooled reversal rates were observed in condition LS ($\mu = 8.4\,\%$), most likely due to access to individual static and dynamic binaural cues (Begault, Wenzel, & Anderson, 2001; Thurlow & Mergener, 1970; McAnally & Martin, 2014). Makous and Middlebrooks (1990) reported similar results in dynamic sound localisation experiments, using bandpass-filtered stimuli between 1.8 kHz and 16 kHz and stimulus durations of 150 ms, reproduced via loudspeakers, which resulted in a 6-% reversal rate. This rate is corroborated by the results from Wenzel et al. (1993), who presented a train of eight 250 ms Gaussion noise bursts, bandpass-filtered stimuli between 200 Hz and 14 kHz, amounting to percentages of 6.5 %.

The reversal rates in condition HP ($\mu = 16.4\,\%$) approximately doubled compared to condition LS. This percentage is still low in relation to headphone-based static localisation studies with non-individual HRTFs (e.g., Wenzel et al., 1993, $\mu = 31\,\%$). Reasons for this performance drop are likely rooted in the combination of mismatched spectral cues in generic HRTFs and the missing access to dynamic cues as provided by head movements, the latter being among the most important factor to reduce the reversal rates (Gilkey & Anderson, 2014;

Oberem et al., 2020). Presenting 3-s long Gaussian noise stimuli via VSSs based on non-individual HRTFs and dynamic reproduction, Wenzel (1995) reported lower front-back but slightly higher back-front reversal rates of 6.7 % and 6.8 %, respectivly, compared to the current study results.

The observed pooled rates of 16.2 % using dynamic binaural reproduction in condition CTC were substantially lower than the ones in specific implementations of static binaural localisation experiments over loudspeakers. Takeuchi and Nelson (2007) tested localisation performance in a CTC system, designed upon the the principle of optimal source distribution (Takeuchi & Nelson, 2002) and installed in an anechoic chamber, with VSSs based on generic HRTFs. As reference stimuli, they used 3 s long pink noise, located at $\varphi = 0°$ in the horizontal plane, followed by a 3-s pause, to present a displaced test VSS, playing back 5 s of pink noise. Possible head movements were prevented by a head rest. Application of this static reproduction setup resulted in front-back and back-front confusion rates of 13.4 % and 15.7 %, respectively. However, also for binaural playback over loudspeakers, there is perceptual evidence that reversal errors can be substantially reduced in interactive binaural auralisation systems compared to static variants. Lentz (2008) conducted perceptual experiments with VSSs reproduced via a 2-loudspeaker CTC system, reproducing two pink noise pulses with a duration of 200 ms each, and a 500-ms inter-stimulus interval. The reversal rate dropped substantially in comparison to static implementation variants, even in a non-ideal reproduction environment with additional reflections from the three intentionally added reflector panels. Although only roughly comparable to the reproduction environment used in the current study, these results again support the positive effect of head movements. It is noteworthy that the reversal pattern observed in condition CTC seemed to be approximately inverted compared to condition HP, with more back-front than front-back confusions.

Monaural cues represent another dominating factor, contributing to localise VSSs in the correct vertical hemisphere (Shaw, 1974; Iida et al., 2007), and affecting the reversal rates when distorted by average characteristics of non-individual HRTFs (Wenzel et al., 1993; Oberem et al., 2020). An even intensified effect is likely observable if the VSSs were generated based on generic HARTFs of behind-the-ear HAs (Kayser et al., 2009; Thiemann & van de Par, 2019; Pausch et al., 2018b) with considerably reduced or even absent pinna-related cues, as done in condition HA. Higher average reversal rates of 41 % and 35 % were reported by Best et al. (2010) in participants with HL fitted with behind-the-ear HAs without and after accommodation, respectively, using monosyllabic speech stimuli up to $f = 4$ kHz. Since the rates were lower in comparison to the completely-in-the-canal HAs, it was suspected that the placement of the behind-the-ear HAs had a decisive effect. It should be added that additional beamforming algorithms in HAs have the potential to decrease the reversal rates (Keidser et al., 2006; Mueller et al., 2012). Apart from distorted monaural cues, potentially influential effects can be attributed to the transducer characteristics of the HA receiver in combination with the open fitting, resulting in strong negative sloping and a corresponding lack of energy towards lower frequencies, hindering access to ITDs. This limited binaural cue access has not only led to increased azimuth localisation errors ϵ_φ, see Figure 5.4, but probably also increased reversal rates. This notion is corroborated by findings by Hebrank and Wright (1974), who observed a decreased localisation performance in the median plane when applying a high-pass filter with increasing cut-off frequency on spatialised white noise stimuli.

In condition CTCwRHA, the pooled reversal rates were reduced ($\mu = 18.9\,\%$). Byrne, Noble, and Glauerdt (1996) attributed a positive effect to localisation cues that are susceptible via residual hearing capabilities when using an open-fit ear piece. Such cues include the previously missing ITDs in condition RHA. In combination with the precedence effect (Gardner, 1968; Litovsky et al., 1999), this improved listening situation had likely led to the improved reversal rates. Given that the pooled reversal rate in condition CTCwRHA only slightly increased compared to condition CTC substantiated the fact that the perceptual contribution of condition CTC dominated the combined playback in condition CTCwRHA.

Overall horizontal source localisation

Binaural listening either relied on individual HRTFs (condition LS), generic HRTFs (conditions HP and CTC), generic HARTFs, or a mixture of both types of directional transfer functions (condition CTCwRHA). The conclusion drawn by Wenzel et al. (1993) that generic HRTFs are sufficient to approximate individual characteristics for a large part of listeners, likely affecting front-back reversal rates, was likewise observed in the current study, since there was no difference in overall horizontal source localisation between conditions HP and LS. This also applied to condition CTC, exhibiting similar performance as observed in condition LS. However, the cue differences in HARTFs compared to HRTFs (Kayser et al., 2009; Pausch et al., 2018b), particularly in ITDs (Pausch, Doma, & Fels, 2021), led to an over- and underestimation of VSS directions in the first and fourth quadrants, respectively. The under- and overestimation in the second and third quadrants were not as pronounced, see Figure 5.3 and Table A.1.

Following up the discussion on the reversal rates, also the horizontal source localisation in condition CTCwRHA seemed to be dominated by condition CTC, overriding the detrimental effects of the HARTFs and allowing comparable performance as in condition LS. Since a fit with open domes only allowed to efficiently convey frequencies between about 810 Hz and 15.4 kHz, considering a dynamic range in magnitude values of 30 dB below the peak value at around 2.6 kHz, see Pausch (2022, Fig. 9b), this improvement may be rooted in the access to additional low-frequency components, provided in condition CTC, and the precedence effect. It needs to be clarified how additional insertion gains and HA algorithms affect the positive influence of the loudspeaker-based binaural playback in condition CTCwRHA.

Angular errors

Complementing the results on the horizontal source localisation performance, a system comparison on the basis of angular error metrics can be considered a refined error analysis, see Figure 5.4 and Table 5.4.

The resulting errors in ϵ_φ, obtained for condition LS, were in the range of the open-loop errors between 1.5 deg and 15.9 deg in the horizontal plane reported by Middlebrooks and Green (1991). Similarly, ϵ_ϑ matched the results by Bronkhorst (1995), with lower values in ϵ_ϑ. In condition HP, ϵ_φ and ϵ_ϑ were in line with the findings of Begault, Wenzel, and Anderson (2001), who observed azimuth and elevation errors of $16.9° \pm 7.8°$ $(\mu \pm \sigma)$ and $17.6° \pm 14.6°$ $(\mu \pm \sigma)$, respectively, presenting VSSs based on generic HRTFs over headphones using dynamic binaural synthesis. A degraded localisation performance using generic HRTFs compared to individual directional transfer functions in condition LS was only measured for ϵ_γ, while the performance in terms of ϵ_φ and ϵ_ϑ did not differ significantly, partially corroborating the conclusions by Bronkhorst (1995), even in the case of dynamic binaural reproduction.

Surprisingly, the results in terms of ϵ_γ, measured in condition CTC, were similar to a mean angle error of 32.4 deg, also representing a great circle central angle, reported by Gardner and Martin (1995) who used static binaural reproduction with VSSs based on generic HRTFs and a CTC system implementation with an upper band-limitation at 6 kHz. Although no effects of the system imperfections discussed earlier were observed in overall horizontal localisation performance, a significant performance decrease appeared to be present for the angular error metrics compared to condition LS. These results motivate further optimisation of binaural playback over loudspeakers using CTC filters if installed in non-ideal reproduction environments (Sæbø, 2001; Kohnen et al., 2016), particularly when the minimum CS was critically undercut, cf. Section 4.4.1 (Parodi & Rubak, 2011; Pausch et al., 2018b).

Significantly increased angular errors were also found in the conditions RHA and CTCwRHA with respect to condition LS. In addition to the underlying reasons, discussed in the context of reversal rates and overall horizontal source localisation, an interesting aspect should be pointed out. The fact that ϵ_φ and ϵ_ϑ between the conditions CTC and CTCwRHA appeared to be very similar suggests that the performance decrease owing to monaural cue distortion is less dominant than the missing low-frequency energy and limited access to ITDs in condition RHA.

In contrast to improved overall horizontal source localisation, no significant effect of additional playback over loudspeakers in condition CTCwRHA compared to condition RHA can be reported regarding angular error metrics. However, trends towards improved performance are clearly visible in ϵ_φ and ϵ_γ.

Limitations

The omission to record the motion tracking data prevented complementary analysis of listener movements and any applied listening strategies (Viveros Munoz, 2019). In this context, it is not possible to unequivocally conclude whether the reduction of the reversal rates in conditions HP, CTC and CTCwRHA in comparison to static binaural reproduction can be attributed to head movements alone. It may also have been the case that the generic HRTF datasets approximated the anthropometric features of some participants to a sufficient extent, which would have helped to lower the confusion rates even if static listening had been applied.

In a real-life scenario, HA users would listen to RSSs, modelled by loudspeakers, as captured and reproduced via the HA microphones and receivers, respectively, applying their individual HARTFs. Mainly because of potential feedback issues, it is not recommended to implement this scenario using the HA prototype for simultaneous recording and playback. Even when applying adequate feedback cancellation algorithms, which alone could bring about undesirable effects, practical problems regarding electrical crosstalk when using unbalanced, slim-diameter cables need to be resolved. However, the application of individual HARTFs with high spatial resolution (Pausch, Doma, & Fels, 2021) further approaches the real-world scenario and also allows to analyse the perceptual effects of generic versus individual directional transfer functions. Such an experiment would additionally enable assessing the perceptual effects of selected HA algorithms on localisation. The significant increase of angular errors in condition CTC motivates further research on binaural reproduction via loudspeakers in non-ideal environments. The critical question whether the combined reproduction approach, implemented in its current form, is sufficiently performant in non-ideal reproduction environments compared to the highly-demanding real-life scenario, remains to be investigated in follow-up experiments.

5.1.4 Conclusion

A dynamic localisation experiment was performed to investigate the effect of different implementation variants of spatial audio reproduction systems. With reference to the localisation accuracy of RSSs, represented by discrete loudspeakers, the perceptual results were compared to the localisation of VSSs, that were binaurally reproduced via headphones, loudspeakers with CTC filters, research HAs alone, and loudspeakers with CTC filters in combination with research HAs. Quite probably for reasons of a limited access to binaural low-frequency cues in HAs with open fit, such as ITDs, reproduction over research HAs alone entailed the highest reversal rates. Using combined reproduction, these missing cues could be supplied via the loudspeakers, effectively lowering the pooled reversal rates. In comparison to static localisation experiments, reduced reversal rates were observed likely due to a dynamic binaural cue access made possible by head movements. Similar localisation performance as observed for RSSs regarding overall horizontal source localisation can be reported in combined reproduction, also entailing a significant positive effect for VSS localisation accuracy compared to the results when using reproduction via research HAs alone. In terms of angular error measures, accounting for VSSs in the horizontal and median planes, the highest accuracy was observed for headphone-based reproduction. The performance gradually decreased for binaural reproduction over loudspeakers with CTC filters, combined reproduction over loudspeakers with CTC filters and research HAs, and research HAs alone. No positive effect of additional loudspeaker-based binaural reproduction on the angular error metrics was found. However, the results regarding ϵ_ϑ indicated that the additional low-frequency cues, provided by loudspeaker playback in combined reproduction, was the prevailing factor for improved localisation performance in comparison to the distorted monaural cues in HARTFs. For future system applications, the localisation results measured in the combined reproduction represent a relevant baseline indicator for the performance of participants fitted with research HAs using an open fitting set to omnidirectional operation mode.

5.2 Experiment 2: Effect of the spatial audio reproduction method on auditory distance perception in simulated room acoustics

In everyday life, auditory distance perception plays an important role for spatial perception. Evaluating the egocentric radial distance of the sound source together with sound source localisation helps to avoid obstacles, especially in situations with poor visibility. It is well understood that acoustic cues such as a decrease in SPL and direct-to-reverberant ratio (DRR), spectral modifications due to dissipation, as well as binaural cues contribute to auditory distance estimates (Kolarik et al., 2015). Oberem, Masiero, and Fels (2016) demonstrated that binaural technology using headphones allows for plausible recreation of VSSs in terms of plausibility detection rates when compared to RSSs. Anderson and Zahorik (2014) tested egocentric auditory (and visual) distance estimation in adults with stimuli relying on 100-ms Gaussian noise bursts that were convolved with a distance-dependent set of measured generic BRIRs. While seated in a hearing booth, the participants were asked to judge the distance of the VSSs, which were presented via headphones. McKeag and McGrath (1996) compared auditory distance perception in HOA ($\mathcal{N} = \{1, 2, 3\}$), virtually reproduced via headphones involving head tracking, to the estimates given RSSs. A comparison between wave field synthesis and VBAP was conducted by Lopez et al. (2014), with varying stimulus types, listening angles and first-order reflections.

The current study

It is unclear how precise egocentric VSS distances can be estimated in simulated room acoustics compared to binaural reproduction via research HAs, loudspeakers with CTC filters, and their combination, based on individual receiver directivities. Experiment 2 aims at evaluating variants of the static hybrid reproduction system, cf. Section 3.4.2, when experienced by HA users. For this purpose, the investigations are extended to test the effect of combined reproduction via research HAs and panning-based reproduction using VBAP and HOA in adults with NH. The experiment applies the 68-channel loudspeaker array installed in the anechoic chamber, cf. Pausch, Behler, and Fels (2020), to answer the following research questions:

Q5.3 How well can the egocentric auditory distance of VSSs in simulated room acoustics be estimated when binaurally reproduced via loudspeakers with CTC filters, research HAs alone, as combined via research HAs and loudspeakers with CTC filters, as combined via research HAs and VBAP, and as combined via research HAs and HOA, in comparison to headphone-based binaural reproduction?

Q5.4 Does the contribution of combined loudspeaker-based playback affect the egocentric auditory distance perception of VSSs compared to its perception using playback via research HAs alone?

5.2.1 Methods

Participants

A different sample compared to the one in Experiment 1 was recruited for Experiment 2. From the 30 adult participants, aged 28.9 ± 6.1 yrs ($\mu \pm \sigma$, range: 22–43 yrs), who participated in the measurements of individual spatial rendering transfer functions with high spatial resolution, see Section 5.2.1, 21 also took part in Experiment 2. Prior to the experiment, a standard ascending PTA with pulsating sine tones for measurement frequencies between 125 Hz and 8 kHz was conducted (Ear 3.0, Hörniß Medizintechnik, Leverkusen, Germany) in the anechoic chamber, see Pausch (2022, Sec. 2.5). To fulfil the pre-defined inclusion criteria, it was necessary to have NH ability, indicated by maximum 4-PTA values of 15 dB and no hearing thresholds exceeding 20 dB. Two participants had only slightly elevated left-ear 4-PTA values of 16.3 dB and 17.3 dB, and were therefore still included. After excluding one participant owing to technical problems and two participants because of exceeding the maximum allowed hearing threshold, the remaining sample included 18 participants (3 female) aged 30.4 ± 6.0 yrs ($\mu \pm \sigma$, range: 22–41 yrs). All participants were informed about their rights and provided written consent. European data protection regulations have been followed to ensure the correct collection, processing and storage of data in a pseudonymous way (European Union, 2016). No representation allowance was paid.

Acoustic measurements in the real-world reference room

The seminar room at the institute with dimensions of $13.8 \times 7.8 \times 2.8 \, \text{m}^3$ (length $\times$ width $\times$ height), resulting in a room volume of approximately $301 \, \text{m}^3$, calculated without furniture, served as reference real-world room for the simulations, see Figure 5.5. For overall acoustic characterisation, room acoustic measurements were conducted according to ISO 3382-2 (2008) at precision level with two source and six receiver positions (three averages per combination), using the same equipment as for the measurements in Pausch (2022, Sec. 2.1.1). Exponentially swept sines with a length of 2^{18} samples were generated at $f_\text{s} = 44.1 \, \text{kHz}$ and used as excitation signal. The mean reverberation times T_{30} were calculated through arithmetic averaging across source-receiver combinations per third-octave band, applying Method C for noise compensation (Guski, 2015), see Figure 5.6a. Given a range of 0.3 to 1.1 s (in third-octave bands), a mean mid-frequency reverberation time of $T_{30,\text{mid}} = 0.7 \, \text{s}$ determines a critical distance of 1.18 m (Kuttruff, 2016).

The simulation required to additionally measure the IRs from a directional loudspeaker (KH 120AW, Georg Neumann GmbH, Berlin, Germany) for the set of distances to be estimated. To obtain a sufficiently high SNR, particularly

Figure 5.5: Visually superimposed distance-dependent IR measurement scenarios in the real-world reference room. The obtained RTs were used to estimate the mean absorption and scattering coefficients of the surface materials for matched RTs in the simulated counterpart. (Photo licensed under CC BY 4.0.)

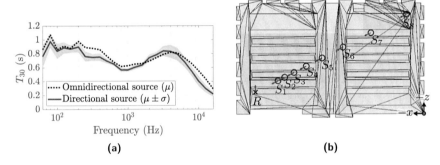

(a)   (b)

Figure 5.6: (a) RTs T_{30} in third-octave bands of the real-world reference room using the dodecahedron loudspeaker, and the directional loudspeaker, sequentially measured for the logarithmically increasing distances d_q. **(b)** Floor plan of the auralised room. The cross and circles indicate the positions of the receiver R and sources S_q at distances $\{d_q | 1 \leq q \leq 8\}$, respectively. View vectors are indicated by red arrows.

for farther distances, each source-receiver combination was measured four times applying exponential sweeps of 2^{18} samples length. The geometrical centre of the loudspeaker and the random incidence microphone (Type 4134 and 2669, Brüel & Kjær, Nærum, Denmark), used with a microphone conditioning amplifier (Type 2690-A, Nexus, Brüel & Kjær, Nærum, Denmark) and an audio interface (RME Madiface XT, Audio AG, Haimhausen, Germany), were both set to a height of

$y = 1.2\,\mathrm{m}$. Given a right-handed global coordinate system with origin in the right rear room corner, cf. Figure 2.1a, the microphone was placed in the left rear area of the room (position: $x = -12.45\,\mathrm{m}$, $z = 1.56\,\mathrm{m}$). These coordinates also defined the position of the virtual listener, cf. Figures 5.5 and 5.6b. A set of $Q = 8$ distances, $\{d_q | 1 \leq q \leq Q\}$ were sequentially measured with respect to the microphone position, at approximately $\varphi = -63.4°$. The distances increased logarithmically, yielding

$$d_q = 10^{\log(d_1) + \frac{\log(d_Q) - \log(d_1)}{Q-1}(q-1)}, \tag{5.2}$$

for the q-th distance, with $d_1 = 1.86\,\mathrm{m}$ and $d_Q = 12.5\,\mathrm{m}$. The final set thus included the distances $\{1.86, 2.44, 3.21, 4.21, 5.52, 7.26, 9.52, 12.5\}$ in meters. Minimum and maximum source distances were chosen so that they surpass the critical distance and the measurement distance of the individual spatial rendering transfer functions presented below, and do not exceed the available horizontal room diameter, respectively. A cord stretched on the ground, a laser distance meter (GLM 40, Bosch Professional, Gerlingen-Schillerhöhe, Germany) and acoustic distances based on the measured IRs were used in combined form to verify the absolute and relative source positions in the room. Figure 5.6a compares the RTs obtained with the directional loudspeaker, averaged across distances, with the results using the dodecahedron loudspeaker. Reasons for the downward deviation in large parts can be attributed to the influence of the directional characteristic during room excitation and mismatched source positions between measurement sessions.

Individual spatial transfer functions

Following the combined reproduction concept, cf. Section 3.4.2, individual rendering and playback receiver directivities are required. It is additionally necessary to use datasets with a high spatial resolution as soon as room acoustic simulations are involved, even in reproduction scenarios with a static listener. The measurement and post-processing steps of these measurements are described below.

Individual spatial rendering transfer functions with high spatial resolution. Ideal individual spatial rendering HRTFs and HARTFs were measured in a separate session using the setup described in Pausch (2022, Sec. 3.9). At the beginning, the participants were equipped with the research HAs and two in-ear microphones (KE3, Sennheiser, Wedemark, Germany). All cables were stuck on neck and clothes to keep the devices in place during the measurement. I used two cross-line lasers to set the ear height, and the front-back and lateral position of the longitudinal

 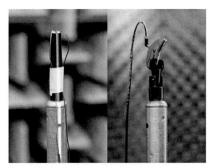

(a) Target pose validation of the participant using two cross-line lasers.

(b) Target pose validation of the microphones during the reference measurements.

Figure 5.7: Positioning of the participant and the reference microphones. (Photos taken from Pausch, Doma, and Fels, 2021, licensed under CC BY-SA 4.0.)

axis of each participant individually, both intersecting with the nominal array centre 2 m above the floor and defining the target pose, cf. Figure 5.7a.

Applying the method of multiple exponential sweeps (Majdak, Balazs, & Laback, 2007; Dietrich, Masiero, & Vorländer, 2013), exponential sweeps with a signal lenght of 2^{15} samples were D/A-converted (RME HDSPe MADI, Audio AG, Haimhausen, Germany) and conditioned by two power amplifiers (MA 32 D, KS Audio Research GmbH, Hettenleidelheim, Germany) to be played back via the loudspeakers. After amplifying and A/D-converting (RME Octamic II, Audio AG, Haimhausen, Germany), the signals picked up by the HA microphones were transmitted to the audio interface (RME Hammerfall DSP Multiface II, Audio AG, Haimhausen, Germany) and processed in MATLAB™.

Since a continuous measurement routine was applied (Richter & Fels, 2019), both the participant and the measurement arc needed to be equipped with tetrahedral rigid bodies and motion tracked. For this purpose, four cameras of the optical motion tracking system, see Pausch (2022, Sec. 3.1), synchronously recorded the trajectories for later post-processing in MATLAB™ (Berzborn et al., 2017). During the measurement, the arc was rotated around the participant who stood still and had to apply continuous self-aligning with the target pose. The current pose error was conveyed by a visual feedback system (Richter, 2019; Pausch, Doma, & Fels, 2021), which was displayed on a 20-inch monitor (SyncMaster 205BW, Samsung, Seoul, South Korea) placed at 2.3 m distance in viewing direction. Favouring more intuitive corrective movements, the pivot point of the participant's head-mounted rigid body was translated individually to coincide with the interaural axis centre. After a training session that aimed at familiarising the participant with the visual feedback system, the actual measurement started. In an initial pause of 20 s, the participant had time for target pose alignment. Thereafter, the arc began to rotate clockwise around the listener, starting from $\varphi = -45$ deg azimuth at a speed of 1.3 deg/s after finished acceleration. This rotation speed allowed a spatial resolution of about 2.49 deg in azimuth angles of the final receiver directivities (Richter, 2019).

For the free-field reference measurements, the same routine was applied, after positioning the in-ear microphones and the research HAs at the nominal array centre, see Figure 5.7b. The HA devices were oriented in target viewing direction, i.e. at $\varphi = 0$ deg, with the horizontal laser beam running through the centre of the microphones to replicate a typical carrying angle.

As a first step during post processing, the raw IRs were time-windowed (right-sided Hann window, 1 ms fade-out starting 2 ms after the latest start instance; left-sided Hann window, 1 ms fade-in 5.8 ms before the end of the fade-out) and cropped to a length of 256 samples. The reference measurements were post-processed likewise. Referencing the time-windowed receiver directivities was accomplished by a regularised spectral division between 100 Hz and 20 kHz and applying a circular shift to obtain a global delay of 1.5 ms. The missing energy in magnitude spectra below 500 Hz due the restricted loudspeaker frequency range was counteracted by extrapolation towards 0 dB at 0 Hz for $f < 500$ Hz, corresponding to physical realities. As a further post-processing step, the resulting transfer functions were directionally mapped, accounting for time-dependent loudspeaker directions and participant pose changes. Global azimuthal rotation was applied to minimise the ITD in HRTF datasets at $\varphi = 0°$ using IACC (Bomhardt, 2017). Finally, the frequency-dependent non-uniform measurement grid was mapped onto a frequency-independent uniform measurement grid of 2.5 deg × 2.5 deg in azimuth and elevation angles to facilitate handling in VR applications (Pausch, Doma, & Fels, 2021). Figures 2.9b and 2.11 show an example set of individual receiver directivities from direction $(\varphi, \vartheta) = (90°, 0°)$.

Individual spatial in-situ playback transfer functions As seen in Figure 3.8, a listener equipped with research HAs without any additional HA algorithms will perceive the reproduced acoustic scene via the ear signals $\check{E}_{\{L,R\}}$ and the HA receiver signals $\check{E}_{HA,\{L,R\}}$. The latter signals result from a convolution of the loudspeaker signals with the front or rear HA microphone signals and subsequent by-ear summation. Therefore, this concept requires to measure the individual long in-situ playback HARTFs $\check{H}'_{\{front,rear\},\{L,R\}}$. For this purpose, each participant was equipped with the research HAs during the pre-measurements prior to the experiment. They were seated on the participant chair with attached head rest in the array centre and instructed to look at the front centre loudspeaker at $\varphi = 0°$ in the horizontal plane. The chair's height and azimuth rotation, as well as the front-rear position of the head rest were set by validating the target pose using two cross-line lasers. While remaining in this pose, sequential IR measurements for each array loudspeaker were conducted using the hardware described in Pausch, Behler, and Fels (2020) with activated FIR filters. The IRs, obtained via exponential sweeps of 2^{15} samples length, were time windowed (Hann window; 1 ms fade-in after 0.5 ms, 5.5 ms fade-out after 250 ms) and cropped. Reference measurements were conducted in the same way but windowed with shorter time window (Hann window; 1 ms fade-in after 0.5 ms, 1 ms fade-out 7 ms after the minimum IR onset time instance), to be inversely convolved with the time-windowed unreferenced signals.

Rendering of binaural and loudspeaker signals

As the basis for the simulation, a simplified geometric model of the real-world reference room was created in SketchUp (Timble Inc., Sunnyvale, California, United States), matching its room volume and exhibiting a total surface area of about $495\,\mathrm{m}^2$. Both sources and receiver were placed in the modelled room, coinciding with the positions during the IR measurements, cf. Section 5.2.1 and Figure 5.6b. For the acoustic simulations, the software RAVEN (Schröder, 2011) was applied, see Section 3.4.1. The simulation aimed at adapting the simulated to the measured RTs T_{30} to be within the relative $\pm 5\,\%$ JND (ISO 3382-1, 2009) for the source-receiver combinations at each distance d_q. Therefore, surface material parameters, i.e. absorption and scattering coefficients, were optimised for each source-receiver combination and subsequently averaged across these combinations to obtain one representative parameter set for the room. To integrate the directivity data of the loudspeaker, measured one-third octave magnitude spectra with $5° \times 5°$ spatial resolution in azimuth and elevation angles, and a length of 512 samples were added to the simulated VSSs. For experimental conditions using binaural reproduction, BRIR or HARRIR filter sets were subsequently created based on the individual spatial rendering transfer functions with high spatial resolution, cf. Section 5.2.1. Panning-based spatial audio reproduction methods, that is VBAP and HOA, required the calculation of loudspeaker weights and spherical wave spectra, respectively. Encoding relied on an SH truncation order of $\mathcal{N} = 7$, see Equation (3.15), which was conducted for all source distances with fixed receiver position. The spherical wave spectra were decoded using an AllRAD+ with max r_E weighting at a decoding order of 7, based on a virtual 8-design (Zotter & Frank, 2019; Politis, 2021a). All room acoustic simulations were executed with third-octave band resolution, an image source order of 2 (early reflections) and

6000 ray tracing particles (late reverberation) in RAVEN. The final filters had a length of 1.5 s, preserving all room acoustical information, cf. Figure 5.6a.

Stimulus

Participants had to estimate the egocentric distances of VSSs after listening to a pink noise pulse with a length of 150 ms and Hann-windowed 2.5 ms on- and offsets, generated in MATLAB™ at $f_s = 44.1$ kHz. The chosen stimulus length enables loudness evaluation at a constant level (Zwicker & Fastl, 2013), which is important for auditory distance perception when mapping and exploiting the decrease in SPL as perceptual cue. Utilising a single pulse should further avoid detrimental echo effects that may occur when playing back pulsed noise trains in indoor environments. Consequently, unfavourable time-staggered reflection patterns may unnecessarily complicate the distance estimation task.

Experimental design and test conditions

The listening experiment was designed as a factorial within-subject design with the predictor *System*, at the levels HP, RHA, CTC, VBAP, HOA, CTCwRHA, VBAPwRHA and HOAwRHA. These factor levels represent the individual and combined reproduction methods with research HAs, predicting the outcome variable *Estimated distance*. The eight test blocks were counterbalanced using a Latin square for avoidance of first-order carry-over effects (Williams, 1949). A continuous visual analogue scale was available to estimate the presented true distance in meter, additionally displaying the estimation as text. The scale was "unlimited" as participants could choose values in a range of 0 to 15 m, where 0 m represented in-head localisation and values above 15 m led to the rating "further away". Both special cases were displayed in the text box. All distances were tested randomly within each block, each distance occurring three times (24 trials). A compulsory break of 30 s was introduced between each block.

Apparatus

As stated above, Experiment 2 was conducted using the 68-channel loudspeaker setup and the related hardware (Pausch, Behler, & Fels, 2020), installed in an anechoic chamber. For headphone-based reproduction, dynamic open-back headphones (HD 650, Sennheiser, Wedemark, Germany) in combination with the described individual headphone equalisation procedure were utilised, cf. Pausch (2022, Sec. 3.3). The key features of the research HAs are summarised in Pausch (2022, Sec. 3.4). While binaural reproduction via loudspeakers relied on the 34-channel subset layout, cf. Figure 4.12, applying a 34-CTC approach with $\beta = 0.01$, the panning-based methods utilised the full layout with the loudspeaker signals weighted as per VBAP weights and HOA decoder. The playback levels in all test conditions were calibrated as described in the upcoming section. The listener sat on the chair with head rest in the centre of the array, which was already set for the measurements of individual in-situ playback HARTFs, cf. Section 5.2.1. For empirical data collection, a tablet with 12.2 in display (Surface Pro 5, Microsoft, Redmond, Washington, USA) was used to display the GUI, and control the experiment computer via Remote Desktop.

Virtual sound sources and calibration

Depending on the experimental condition, either the BRIR or front HARRIRs filter sets, or the simulated loudspeaker signals were linearly convolved with the stimulus for VSS generation. In all combined reproduction scenarios, that is CTCwRHA, VBAPwRHA and HOAwRHA, the binaural signals of the research HAs were additionally convolved with the previously measured individual front playback HARTFs. The HA signals featured a delay of $\Delta t_{\mathrm{HA}} = 5\,\mathrm{ms}$, relative to the corresponding loudspeaker-based signals, which was verified through evaluating the combined IR measurements, cf. Section 4.3.2. No additional HA algorithms were included.

To account for the substantial spectral differences across reproduction methods, e.g. between the conditions HOA and RHA, a loudness-based calibration of the reproduction level was chosen. Defining condition HP as reference, I evaluated the signal of the right ear simulator of the artificial head, cf. Pausch (2022, Sec. 3.2), to obtain a loudness about 11.3 sone, corresponding to approximately 70 dB SPL, for the closest VSS distance (DIN 45631, 1991; Berzborn et al., 2017). The remaining individual conditions were calibrated analogously by setting the gains g_{LS} and g_{HA}, cf. Figure 3.8. In combined reproduction scenarios, the overall level was likewise adjusted by g_{global} to achieve the same combined total loudness as in condition HP. As an example, the specific loudness and overall loudness results for the individual playback paths in condition HOAwRHA are shown in Figure 5.8 .

Experimental procedure

After collecting the personal data and signed informed consent and providing information about the general experimental procedure, the participants were informed about their task to estimate the egocentric distance of a sound source situated in a room with simulated room acoustics. Thereafter, the participants were seated on the participant chair, located in the array centre to conduct the PTA. In case of inconspicuous results, the experiment continued with measuring individual headphone transfer functions and – after adjusting the height

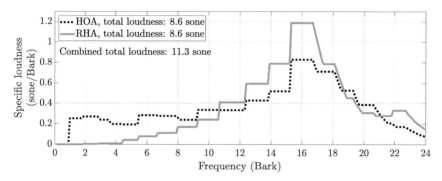

Figure 5.8: Loudness-based calibration of the playback levels, shown by example for condition HOAwRHA.

of the chair and head rest – the individual in-situ front playback HARTFs. Before the first test block, the participants were given precise explanations of the geometrical property *distance*, according to Lindau et al. (2014), with graphical representation. Participants were asked to listen to the stimuli with closed eyes. The question displayed above the visual analogue scale was: "How far away is the sound source?". No training or external distance reference was provided and each stimulus could only be listened to once. The whole experiment took about 70 ± 16 min ($\mu \pm \sigma$).

Egocentric distance model

According to Zahorik (2002), egocentric distance estimates can be well approximated using a compressive power function of the form

$$\hat{d} = \xi d^{\zeta} \quad \text{(m)},\tag{5.3}$$

with $\hat{d}$ and d symbolising estimated and true distances, respectively. The model parameter ξ describes the amount of linear compression or expansion, whereas ζ accounts for non-linear compression or expansion. This model is simplified to a linear one if both true and estimated sound source distances are analysed on double-logarithmic axes with base-10 logarithms, yielding

$$\log(\hat{d}) = \log(\xi) + \zeta \log(d).\tag{5.4}$$

Ordinate intercept and slope are accordingly defined by $\log(\xi)$ and ζ, respectively.

Table 5.6: Hypotheses of Experiment 2.

Code	Hypothesis
H1-Exp2	In comparison to auditory distance estimates of VSSs given headphone-based reproduction, the estimates differ in terms of model parameters and RMS errors for VSSs based on HA-based and loudspeaker-based rendering and reproduction methods, and the combination thereof.
H2-Exp2	Auditory distance estimates differ in terms of model parameters and RMS errors for VSSs rendered and reproduced as combined via HA-based and loudspeaker-based rendering and reproduction methods, compared to when rendered and reproduced via research HAs alone.

Auditory horizon

Zahorik (2002) also reported that humans overestimate only close sound sources whereas most source distances are underestimated considerably. The crossover point Ξ between over- and underestimation is referred to as auditory horizon, or specific distance tendency (Gogel, 1969), and can be calculated as

$$\Xi = 10^{\frac{\log(\xi)}{1-\zeta}} \quad (m). \tag{5.5}$$

Error analysis

In addition to a linear regression model evaluation, the RMS errors between estimated and true distances, $\{\hat{d}_q\}$ and $\{d_q\}$, respectively, were calculated. For hypothesis testing, averages across distances per experimental condition shall serve as global error metric.

Hypotheses

The initial research questions resulted in two hypotheses, H1-Exp2 and H2-Exp2, as summarised in Table 5.6.

5.2.2 Results

In a first step, the data estimates were averaged over the three trials per distance within the reproduction systems to avoid pseudo replication. For increased robustness in estimates during the regression analyses described below, outliers above and below 1.5 times the upper and lower interquartile ranges, respectively, were removed from the averaged data per distance and condition (3.6 % outliers). While no in-head localisations were perceived by the participants, about 0.6 %

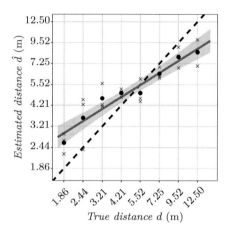

Figure 5.9: Example linear model fit for the results measured for one participant in
condition RHA. While the dashed diagonal line represents an optimal fit
($\xi = 1$, $\zeta = 1$), dots show distance estimates, averaged across the three by-
participant estimates per distance (crosses), and fitted by the solid regression
line ($\xi = 2.05$, $\zeta = 0.58$, $R^2 = .82$) with 95 % confidence region. The resulting
auditory horizon lies at $\Xi = 5.45$ m.

of all presented distances were rated as being "further away", and subsequently
also removed prior to the analysis.

Model coefficients and auditory horizon

The model introduced in Equation (5.4) was evaluated by participant over the
levels of the factor *True distance* for each experimental condition, i.e. levels of
System, resulting in 18 individual parameter sets for ξ, ζ and R^2 per condition.
As an example, the model fit for one participant in condition RHA is shown
in Figure 5.9. Note that both presented and true distances are plotted on logarithmic axes, entailing equidistant distance increments due to their logarithmic
spacing. Figure 5.10 presents the corresponding model parameters, in comparison
to previous modelling results from Zahorik, Brungart, and Bronkhorst (2005) and
Anderson and Zahorik (2014). Owing to the high inter-participant variability, the
resulting auditory horizons were cleaned from outliers as done during data pre-
processing (15.3 % outliers). No value Ξ was calculated for the comparison values
since ξ and ζ represent averaged inter-study and inter-participant results, respectively, which are not meaningful when used in combination. By-condition mean
and standard deviations of model parameters and resulting auditory horizons,
averaged across participants, are presented in Table 5.7.

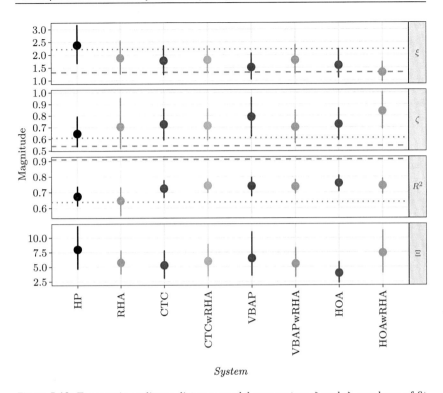

Figure 5.10: Egocentric auditory distance model parameters ξ and ζ, goodness-of-fit parameter R^2, and auditory horizon Ξ, per condition in Experiment 2. Dots and error bars indicate means and standard deviations, $\mu \pm \sigma$. Horizontal dashed and dotted grey lines depict the model parameter values reported by Zahorik, Brungart, and Bronkhorst (2005) and Anderson and Zahorik (2014, auditory condition "A"), respectively.

To investigate the effect of the within-participant factor *System* on ξ and ζ, two repeated-measures ANOVA with Type III sum of squares were conducted. Prior to the analysis, all data points associated with ξ and ζ were transformed by adding a constant value of 1, subtracting the respective minimum of all data points, and subsequently taking the square root. Assumptions of normality were checked running Shapiro-Wilk tests on transformed data of all data cells by experimental condition. Test results suggested that 75 % and 87.5 % of residuals related to ξ and ζ come from a normal distribution. As normality assumptions were widely met (Pearson, 1931; Schmider et al., 2010), the parametric analysis was maintained and interpreted at a confidence level of 95 %. Planned comparisons

Table 5.7: Summary of egocentric distance model coefficients ξ, ζ, and R^2, as well as resulting outlier-free auditory horizons Ξ, each averaged across participants.

Author(s)	System	Model coefficient			
		ξ $\mu \pm \sigma$	ζ $\mu \pm \sigma$	R^2 $\mu \pm \sigma$	Ξ (m) $\mu \pm \sigma$
	HP	2.38 ± 1.69	0.64 ± 0.29	0.67 ± 0.13	$8 \quad \pm 7.5$
	RHA	1.88 ± 1.42	0.70 ± 0.49	0.65 ± 0.19	5.74 ± 4.08
	CTC	1.77 ± 1.21	0.73 ± 0.30	0.72 ± 0.12	5.29 ± 4.66
Pausch (2021)	CTCwRHA	$1.8 \quad \pm 1.1$	0.71 ± 0.29	0.74 ± 0.10	5.98 ± 5.53
	VBAP	1.52 ± 1.11	0.79 ± 0.36	0.74 ± 0.13	6.49 ± 7.64
	VBAPwRHA	1.80 ± 1.22	0.70 ± 0.33	0.74 ± 0.10	5.53 ± 5.03
	HOA	1.59 ± 1.23	0.73 ± 0.33	0.76 ± 0.12	3.94 ± 3.47
	HOAwRHA	1.32 ± 0.89	0.84 ± 0.33	0.74 ± 0.11	7.39 ± 7.72
	Grand $\mu \pm \sigma$	1.76 ± 1.26	0.73 ± 0.34	0.72 ± 0.13	$6 \quad \pm 5.83$
Zahorik, Brungart, and Bronkhorst (2005)	Various	1.32 ± 0.75	0.54 ± 0.21	0.91 ± 0.13	
Anderson and Zahorik (2014)	(HP)	2.22 ± 1.99	0.61 ± 0.30	0.64 ± 0.22	

as required for statistical hypothesis testing were conducted on the transformed data using multiple t-tests with Holm–Bonferroni correction (Holm, 1979).

An ANOVA revealed a significant main effect of *System* on ξ, $F(3.53, 59.94) = 7.88$, $p = .003$, $\eta_p^2 = .23$. Due to violated sphericity assumptions according to Mauchly's test results, $\chi^2(27) = 56.43$, $p = .001$, Greenhouse-Geisser estimates of sphericity, $\varepsilon = .5$, were applied to correct the degrees of freedom. Test results from planned comparisons indicated significant decreases in ξ, comparing RHA versus HP, $t(119) = -2.52$, $p = .021$, CTC versus HP, $t(119) = -2.88$, $p = .019$, CTCwRHA versus HP, $t(119) = -2.61$, $p = .021$, VBAP versus HP, $t(119) = -4.29$, $p < .001$, VBAPwRHA versus HP, $t(119) = -2.74$, $p = .021$, HOA versus HP, $t(119) = -3.91$, $p < .001$, HOAwRHA versus HP, $t(119) = -5.3$, $p < .001$. Additionally, a significant decrease between HOAwRHA versus RHA was present, $t(95) = 13.84$, $p < .001$. These results support hypotheses H1-Exp2 and H2-Exp2 regarding differences in ordinate intercept ξ, and perceptual effects of HOA in combined reproduction HOAwRHA compared to reproduction via research HAs alone (condition RHA), respectively.

An ANOVA indicated no significant main effect of *System* on the slope ζ, $p > .05$, thus not supporting hypotheses H1-Exp2 and H2-Exp2 regarding ζ.

Allowing pairwise instead of listwise deletion of missing data points due to outlier removal, the effect of *System* on Ξ was investigated on the basis of an

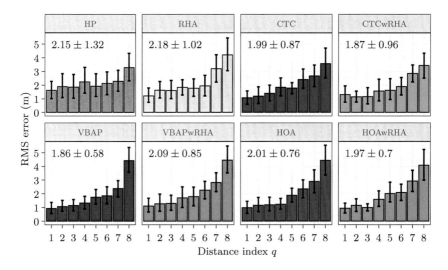

Figure 5.11: Distance-dependent RMS error split into the levels of *System*. Bars and error bars represent means and standard deviations, $\mu \pm \sigma$, respectively, averaged across participants. Provided values represent $\mu \pm \sigma$, averaged across participants and distances.

LME model with the same optimisation method as used for Experiment 1 (Bates et al., 2015). *System* was entered as fixed effect, while participant responses were modelled as random intercepts. Follow-up Type II Wald χ^2 tests indicated no significant effect, $\chi^2(7) = 12.24$, $p = 0.093$, thus not supporting hypotheses H1-Exp2 and H2-Exp2 regarding Ξ.

Error analysis

Figure 5.11 presents the by-condition distance-dependent RMS errors, averaged across participants, showing increasing tendencies towards farther distances in all experimental conditions. The effect of the within-participant factor *System* on the distance-averaged RMS error was evaluated by means of a repeated-measures ANOVA with Type III sum of squares. Shapiro-Wilk test results indicated that 87.5 % of the residuals of square root transformed data cells by experimental condition were normally distributed. The omnibus analysis revealed no significant main effect of *System* on transformed RMS errors, $F < 1$, $p > .05$, therefore not supporting hypotheses H1-Exp2 and H2-Exp2 regarding global distance estimation errors.

5.2.3 Discussion

The frequently reported over- and underestimation of distances can be confirmed once again by the current results. Model parameters obtained for the reference condition HP were largely in line with those reported by Anderson and Zahorik (2014, auditory condition "A": $\xi = 2.22$, $\zeta = 0.61$), who presented VSSs over headphones based on measured generic BRIRs of a room with a broadband RT of 1.9 s. Slightly higher R^2 values were observed, corroborating the effectiveness of the underlying auditory model. Although to be viewed critically, for aforementioned reasons, a farther auditory horizon of $\Xi = 4.44$ m compared to the previously reported $\Xi = 3.23$ m can be calculated based on median values for ξ and ζ. However, Ξ was substantially pushed upwards when calculated based on individual model parameters, cf. Figure 5.10 and Table 5.7.

These results suggest that the room acoustic simulations were capable of replicating the previously listed perceptual distance cues to a sufficient extent when reproduced via headphones, although resulting in farther distance estimates compared to the literature results. Given the lower mean reverberation times in the simulated VAE of the current study, lower values for Ξ would have been expected, since longer RTs typically result in larger auditory horizons (Mershon & King, 1975). This said and given the loudness calibration, reasons for overestimation are likely caused by discrepancies in distance cues. For a follow-up analysis, the distance-dependent BRIRs were measured from the custom-made artificial head, see Pausch (2022, Sec. 3.2). The artificial head was placed at the same position as the random-incidence microphone during the IR measurements in the real-world reference room, oriented in nominal viewing direction to $\varphi = 0°$, cf. Figure 5.6b. For the measurement, the same equipment and settings were used. Comparisons between measured and simulated binaural signals indeed revealed well-matched relative SPL decays but lower DRRs throughout all distances in the simulations. Differences in DRRs are visualised for condition HP in Figure 5.12. These results were obtained by calculating the energy ratio in dB between the DS and the reverberant part, i.e. the remaining BRIR. The DS was extracted using a Hann window starting 0.5 ms before and ending with a 0.5 ms fade-out 2 ms after the BRIR start of the corresponding ear signal. Consequently, the reverberant part was obtained by applying a left-sided Hann window with 0.5 ms fade-in after the end of the DS (Albrecht & Lokki, 2013). The calculations were carried out for the measured (abbreviated as "meas") distance-dependent BRIRs, and the simulated (abbreviated as "sim") BRIRs, averaged across participants per distance. Both left and right simulated BRIRs exhibit substantially lower DRRs. Although this result varies, the tendency to underrepresent DRRs was present in all conditions, thus likely being a main contributor for increased auditory hori-

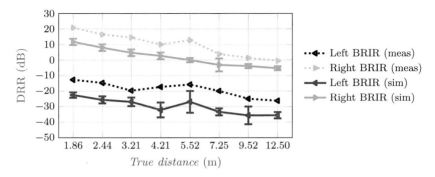

Figure 5.12: Comparison of DRRs between measured BRIRs, and simulated BRIRs in condition HP. Triangles and error bars represent means and standard deviations, $\mu \pm \sigma$, respectively.

zons Ξ. To be fair, it must be added that the underlying optimisation routine in RAVEN (Schröder, 2011) only aimed to match the RTs in third-octave bands by optimising the absorption and scattering coefficients of the wall materials, without attempting to match the temporal energy distribution of the reference IRs.

The ordinate intercepts ξ in the remaining conditions were significantly lower, compared to condition HP, lying between the comparison values by Zahorik, Brungart, and Bronkhorst (2005, $\xi = 1.32$), and Anderson and Zahorik (2014, $\xi = 2.22$), the latter authors reporting data from a power fit function analysis with varying experimental factors. Also, the corresponding slopes ζ tended to increase, without reaching significance, with consistently higher R^2 values in individual and combined loudspeaker-based conditions. Although exhibiting the lowest mean R^2 values and a pronounced spread in ζ, auditory distance perception in condition RHA is comparable to a large part of the remaining loudspeaker-based methods, when used individually and in combination. It appears that the distance cues were still effectively conveyed despite differences in HARTFs compared to HRTFs, together with the limited frequency range and other spectral characteristics of the research HA receivers, cf. Pausch (2022, Fig. 9b). McKeag and McGrath (1996), who compared auditory distance perception in HOA ($\mathcal{N} = \{1, 2, 3\}$), virtually reproduced via headphones with head tracking, to the estimates given RSSs, concluded that first-order binaural Ambisonics downmixes already provides sufficient auditory information for distance information. In this study, the contribution of HOA in HOAwRHA seemed to have had an effect on combined distance perception, significantly decreasing ξ and tending to increase ζ, without having an effect on Ξ. Calculating the auditory horizon Ξ based on grand median values of model parameters across all experimental conditions ($\xi = 1.52$, $\zeta = 0.65$), resulted in $\Xi = 3.29\,\text{m}$, which again should be interpreted with caution. Although the RMS error gradually increased with distance, overall distance-averaged results suggested comparable performance among the participants regardless of the reproduction method used.

5.2.4 Conclusion

Egocentric auditory distance estimation was assessed across different implementation variants of the proposed combined static hybrid auralisation system. Participants were presented VSSs at logarithmically spaced distances, placed in the equivalent VAE of a real-world seminar room with matched RTs. A compressive power function was used to fit the distance estimates obtained for signals binaurally reproduced via headphones, research HAs, and loudspeakers with CTC filters, all based on individual datasets of spatial rendering transfer functions with high spatial resolution. Loudspeaker-based reproduction also included VBAP and HOA, which were tested in combination with research HAs, including short individual in-situ playback transfer functions. Previously reported results for headphone-based reproduction were largely replicated, except for higher specific distance tendencies. The under-representation of DRRs in the simulations was identified as the potential main reason for this increase. Participants performed similarly as observed in the loudspeaker-based methods when estimating distances via research HAs and in combination. An effect of loudspeaker-based methods on auditory model intercept ξ was only observed for HOA in combined reproduction with research HAs, compared to reproduction via research HAs alone. No difference in performance was found across conditions in terms of RMS errors, when averaged across distances. Overall, these result suggest less salient perceptual effects of the position and electroacoustic characteristics of the research HAs as they occurred during localisation of VSSs, already in free-field conditions.

6

System application

Experiment 3: Speech perception of children with hearing aids in simulated room acoustics

One of the various possibilities of using the combined binaural real-time auralisation system, introduced in Chapter 3, will be demonstrated within the framework of a clinical experiment. Speech intelligibility of a target talker in the presence of two competing distractors was investigated in children with symmetric bilateral HL, using an established binaural SRM paradigm based on a sentence recognition task, which was implemented in a virtual classroom with simulated room acoustics. The children were bilaterally fitted with research HAs, which were operated in omnidirectional mode, and received frequency-dependent insertion gains derived from their individual audiograms. A subset of children with HL wearing their own individually fitted commercial HAs underwent the same test procedure, representing an important benchmark group to confirm the general validity of the HA simulation. The results obtained for the control group, consisting of age-matched children with NH, allowed comparing between-group effects resulting from the systematic manipulations of experimental factors.

6.1 Introduction

Classrooms are one of the most important learning environments for children over a long period of growing up. Not limited to, but especially in classical frontal teaching, a crucial task is to focus on the information conveyed by the teacher

Related publications:

Peng, Pausch, and Fels, 2016a; Pausch et al., 2016c; Bell et al., 2020; Peng, Pausch, and Fels, 2021.

while ignoring competing noises. Noise types include energetic maskers, generated by external sources outside the room, internal operating noises from electronic devices (e.g. projectors), heating, ventilation and air condition systems (DIN 18041, 2016), which decrease the overall SNR, and may have an impact on academic success (Shield & Dockrell, 2008). Informational maskers, such as distracting talk from fellow students, dominantly affect SiN perception performance (Brungart, 2001). It is known from a number of studies that children perform worse in multi-talker scenarios compared to adults when tested for their ability to extract the target information (e.g. Hall III et al., 2002; Wightman & Kistler, 2005b), manifesting itself as an upward shift in SRTs, cf. Section 2.9. More vulnerable groups, including children with HL, particularly struggle to decode relevant information (Torkildsen et al., 2019; Bronkhorst, 2000), owing to the limiting factors presented in Section 2.6.3, which is aggravated by adverse room acoustics (Iglehart, 2020).

Selectively directing one's attention on a target talker to perceive the conveyed information is supported when the distracting talkers are spatially separated from, as opposed to spatially co-located with, the target source – an advantage that is referred to as SRM (Ching et al., 2011; Misurelli & Litovsky, 2015), cf. Section 2.9. Even in scenarios with symmetrically displaced distractors, where effects of "better monaural ear" listening and exploitation of a head shadow effect are absent (see, e.g., Hawley, Litovsky, & Culling, 2004; Marrone, Mason, & Kidd, 2008), binaural and monaural cues, cf. Section 2.8.1, can still be applied for the purposes of sound source localisation and auditory perceptual grouping (Bregman, 1994). In children, the ability of segregating the target stream develops in early to mid childhood (Litovsky, 2005; Yuen & Yuan, 2014; Cameron & Dillon, 2007; Cameron, Glyde, & Dillon, 2011; Peng, Pausch, & Fels, 2021). There is evidence that young adults with HL cannot benefit to the same extent from spatially separated distractors as listeners with NH (Marrone, Mason, & Kidd, 2008). It should be clarified in the current study if a similar effect can be observed in young children with HL and if there is a specific effect of reverberation.

In connection to investigating the impact of informational maskers on SRTs and related SRM, it is of particular interest whether a specific advantage can be measured when the distractors do not share the same voice as the target talker. The available spectral difference between sources was identified as being a relevant cue for unmasking speech (Bronkhorst, 2000). Cameron and Dillon (2007) reported that already 5-yr-olds make use of voice differences to improve their SiN perception performance, although the positive effect is less pronounced compared to SRM. With reference to the temporally fluctuating envelope of speech, it stands to reason to extract and extrapolate target contents from sections that are less strongly masked via temporal glimpsing (see, e.g., Vestergaard, Fyson,

& Patterson, 2011), which is particularly useful given speech masked by single-talker distractors. However, owing to a reduced temporal resolution and elevated high-frequency hearing thresholds, this ability is usually impaired in listeners with HL (Rana & Buchholz, 2018).

Crandell and Smaldino (2000) discusses various acoustic variables within the scope of classroom acoustics that may negatively affect speech perception in children with NH and HL. A notable portion can be attributed to reflections from surfaces, determining the RTs of a room, cf. Section 2.3.2. Even though the superposition of early reflections with the DS favours increased loudness and speech intelligibility (Thiele, 1953; Seraphim, 1961), too high a proportion of late reverberation will degrade consonant perception by temporal smearing. This is especially true when consonants follow vowels, which act as a temporal masker, the effect of which is likely reinforced by adverse room acoustics and the upward spread of masking (Moore, 2012). Further detrimental consequences of reverberation can be derived for binaural cues insofar as reflections tend to reduce ILDs and hamper the evaluation of ITDs by lowering IACC (Rakerd & Hartmann, 2010). Children with sensorineural HL also require a substantially higher SNR to be able to parse the conveyed information as effectively as their peers with NH (Crandell & Smaldino, 2007). This fact motivated standards to adapt recommendations of target RTs in core learning spaces (ANSI/ASA S12.60-2010/Part 1, 2010) and classrooms for inclusive education (DIN 18041, 2016).

The extent to which the aforementioned factors impact SiN perception of children with symmetrical HL, bilaterally fitted with research HAs, compared to age-matched children with NH shall be investigated under controlled laboratory conditions, using the potential of VAEs and state-of-the-art interactive real-time auralisation combined with room acoustic simulations. In particular, Experiment 3 aimed at answering the research questions outlined below.

Q6.1 How does reverberation affect SRTs? Are there performance differences in children with HL and children with NH?

Q6.2 Does performance in SiN perception change differently under varying RTs when the distractors use different voices than that of the target voice? Are there performance differences in children with HL and children with NH?

Q6.3 Does SRM improve differently under varying RTs? Are there differences in children with HL and children with NH?

Q6.4 Is the HA simulation capable of replicating the SRT results measured for children wearing their own commercial HAs?

6.2 Methods

6.2.1 Participants

Twenty children (11 female) with HL aged 10.5 ± 2.0 yrs ($\mu \pm \sigma$, range: 7.2–13.4 yrs), all German native speakers, participated in the study. The children were recruited through teacher-parent communications from the David-Hirsch-Schule in Aachen, a special school for children with HL focusing on hearing and communication, and from the patient database of University Hospital Aachen. After analysing the hearing thresholds, measured as per standard ascending PTA (ISO 8253-1, 2010) between 250 Hz and 8 kHz, I removed three children from the dataset, one suffering from unilateral HL, the second one from unilateral deafness, and the last one having articulation disorder (type F80.2[1]), attention deficit hyperactivity disorder (type F90.0[1]), and trouble falling asleep with accompanying exhaustion (type F84.8[1]). The remaining 17 children (10 female) aged 10.3 ± 2.0 yrs ($\mu \pm \sigma$, range: 7.2–13.3 yrs) had bilateral HL, were bilaterally fitted with HAs, and their parents reported no history of cognitive or neurological disorders and language deficits. Individual and averaged air-conduction PTAs of the sample are shown in Figure A.1. Mean 4-PTAs averaged across 0.5, 1, 2 and 3 kHz, provided as part of the demographic information in Table A.2, resulted in a pooled average of 63.7 ± 22.6 dB HL ($\mu \pm \sigma$, range: 11.3–105 dB HL) and 65.7 ± 23.3 dB HL ($\mu \pm \sigma$, range: 16.3–100 dB HL) for the left and right ear, respectively. Average symmetric PTA thresholds of -2 ± 9.6 dB HL ($\mu \pm \sigma$, range: -31.3–16.3 dB HL) were calculated as differences in 4-PTAs across ears. An analysis of audiometric slopes, evaluated through linear regression over frequencies of 0.5, 1, 2 and 3 kHz, further revealed flat or gradually negatively and positively sloping audiograms, cf. Figure A.1 and Table A.2.

The age-matched control group consisted of 28 children (10 female), all German native speakers, and were aged 10.1 ± 1.6 yrs ($\mu \pm \sigma$, range: 6.9–12.4 yrs). All children with NH passed the legally obligatory school entry health examination (Schulgesetz NRW - SchulG, 2021), conducted by pediatricians or general practitioners, including, among others, tests on number and set theory comprehension, an eyesight test and headphone-based PTA. The children's parents additionally reported no history of HL, cognitive and neuronal disorders, or language impairment.

Ethical study permit was granted by the ethical committee of the University Hospital Aachen, Germany (EK 188/15). The study was conducted in compliance to the World Medical Association Declaration of Helsinki (General Assembly of

---

[1]Code as per ICD-10-GM, 2020.

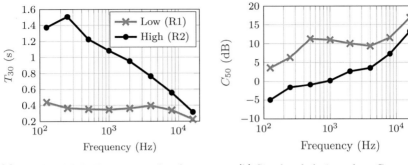

(a) Simulated RTs T_{30} in octave bands, representing the levels Low and High of the factor *Reverb*, coded as R1 and R2, respectively. The legend also applies to sub-figure **(b)**.

(b) Simulated clarity values C_{50}.

Figure 6.1: Room acoustic parameters of the two simulated classroom conditions.

the World Medical Association and others, 2014). All personal data and experimental results collected were processed and archived according to country-specific data protection regulations. Prior to the experiment, children and their parents were informed of the rights to which they are entitled and subsequently provided assent and consent, respectively. The children were reimbursed for participation.

6.2.2 Virtual acoustic environment

Simulated classroom

The experiment was conducted in a virtual classroom, the room acoustics of which were simulated based on the computer-aided design model of a cuboid room with dimensions 11.8 m×7.6 m×3 m (length × width × height), resulting in a room volume of approximately 244 m³, see Peng et al. (2016). The model featured typical furniture with geometry details simplified for simulation (97 polygons, 34 planes, surface area of 383 m²) in RAVEN (Schröder, 2011), see Chapter 3, Section 3.4.1. All simulations described below were conducted using an image source order of 2 and 20,000 ray-tracing particles at 20 °C and 50 % relative humidity. By application of the same optimisation routine as used in Experiment 2, cf. Chapter 5, Section 5.2.1, the absorption and scattering coefficients of the surface materials were adapted in a way to obtain the two RTs presented in Figure 6.1a, representing two acoustic versions of the same classroom. While the choice of the lower RTs was motivated by DIN 18041 (2016, p. 12, Eq. 4), suggesting a target RT of 0.48 s for the given volume when using the room for inclusive education, the higher ones aimed at reflecting challenging acoustic conditions, as they would occur in poorly optimised seminar rooms or larger lecture halls. The mean mid-frequency RTs in the respective simulated condition amounted to $T_{30,\text{mid}} = 0.4\,\text{s}$ (low RT) and $T_{30,\text{mid}} = 1.2\,\text{s}$ (high RT). Ranging from $3.5\,\text{dB}$ to $17\,\text{dB}$ (low RT) and $-1\,\text{dB}$ to $13\,\text{dB}$ (high RT), the corresponding clarity indices C_{50}, see Figure 6.1b, resulted in intelligibility-weighted and summed composite values C'_{50} of about $10.3\,\text{dB}$ (indicating excellent speech intelligibility) and $1.7\,\text{dB}$ (indicating fair speech intelligibility), respectively (Marshall, 1994). As simulated listening scenario, the virtual listener was located in the front row of the classroom with VSS positions as per experimental paradigm, see Section 6.2.4, exhibiting egocentric distances of $2\,\text{m}$ each.

Simulation of binaural ear signals

The simulation of the DS in the BRIRs, used for the external sound field simulation in both participant groups, relied on individualised sets of the generic HRTF datasets, see Chapter 4, Section 4.2.1, with a reduced spatial resolution of $3° \times 3°$ in azimuth and elevation angles. Due to the fact that the anthropometrics in children substantially differ from that of adults (Clifton et al., 1988; Fels, Buthmann, & Vorländer, 2004), the individualisation procedure involved scaling of the ITDs (Bomhardt, de la Fuente Klein, & Fels, 2016; Bomhardt, 2017) and spectral features (Middlebrooks, 1999). Three representative head sizes were derived based on a collection of head dimensions (depth, height and breadth) in children (Fels, 2008), representing the 15th, 50th and 85th percentiles. Prior to the experiment, the head dimensions were measured from each child using a caliper gauge, to subsequently select the best-fit dataset from the pre-calculated database. The individualised dataset was used for rendering and reproduction purposes of the external sound field, see Figure 3.7.

Simulation of binaural hearing aid signals

The individualisation of the front HARTFs to be used for simulating the DS of the HA signals followed the same procedure as described above, except that the spectral cues were not adjusted since the underlying algorithm by Middlebrooks (1999) is only meaningful for HRTFs, and HARTFs do not contain the same spectral characteristics, cf. Section 2.8.2. Based on the individual audiograms of the participants with HL, see Figure A.1, frequency-dependent insertion gains as per Moore et al. (1999) were calculated on the MHA (Grimm et al., 2006), accordingly amplifying the HA receiver signals. Although a closed fitting would be advisable for some of the measured audiograms, for comparability reasons and facilitated external sound field perception, all participants were fitted using the open-dome ear piece, whose passive damping properties are presented in Pausch (2022, Fig. 8a). A restriction of maximum permissible peak levels to 105 dB SPL by means of software and hardware limiting prevented overamplification and harmful exposure during HA-based playback. No other HA algorithms were utilised, imitating open-fit HAs operated in omnidirectional mode.

6.2.3 Apparatus

For Experiment 3, the hardware required for the specific implementation of the binaural real-time auralisation system, cf. Section 3.4.1, was installed in a custom-made hearing booth, cf. Pausch (2022, Sec. 2.1 and 3.5). All signals for binaural reproduction via the research HAs and loudspeakers were conditioned and levelled using an audio interface (RME Fireface UC, Audio AG, Haimhausen, Germany). Loudspeaker-based binaural reproduction relied on 4-CTC (Sanches Masiero, 2012), cf. Section 3.3.3. The simulated HA signals were processed on the MHA, as described above, and played back via the HA receivers with a delay of $\Delta t_{HA} = 5$ ms, relative to loudspeaker-based reproduction. Participant IDs marked with asterisks in Table A.2 felt uncomfortable wearing the unfamiliar research HAs in combination with the fitting provided by the MHA and thus used their own pair of HAs to listen to the simulated scenes. For reasons of uncontrollable configuration parameters and to avoid a bias in results, these children were therefore excluded from the core analysis, plotted as comparison data points, and analysed separately. The elimination of the HA-related path in the system made it necessary to adopt rendering and auralisation by simulating the signals at the microphone positions of the commercial HAs via binaural loudspeaker reproduction. Specifically, the 4-CTC filters were calculated based on the front rendering HARTFs with high spatial resolution, cf. Section 4.2.1. The resulting inaccuracies in sound field perception via the blocked ear canals were deliberately accepted in the comparative sample.

For calibration purposes, speech-shaped noise, matching the combined long-term average speech spectrum of the voices of target talker and distractors, was played back via the VSSs, as required in the different test conditions, in both reverberant conditions, see Figure 6.2. After placing the artificial head without ear simulator, see Pausch (2022, Sec. 3.2), at the nominal listening position in the centre of the hearing booth, the across-ear RMS values of the ear signals were matched across all experimental conditions, allowing to compare SNRs and measured SRTs. To calibrate the reproduction levels for the group of children with HL, the research HAs were sequentially attached to the artificial head with ear simulator, see Pausch (2022, Sec. 3.2), to measure the RMS in HA-based playback without any fitting involved (flat audiogram). Consequently, the path-dependent gains, see Figure 3.8, were set to obtain matched individual reproduction levels. Their combined level was subsequently corrected by the combined gain factor, cf. calibration procedure described in Section 5.1.1, to match the playback level set for children with NH.

The children were seated on a chair at the centre of the hearing booth, i.e. the centre of the loudspeaker array. To enable interactive auralisation, an optical motion tracking system with four cameras, see Pausch (2022, Sec. 3.1), continuously captured their head pose and updated the DS of either the HRIRs alone or combined with the HARRIRs in real time, as per Configuration A of Table 3.1. For correct auralisation, the offset of the head-mounted rigid body to the centre of the interaural axis was corrected before the experiment (Berzborn et al., 2017). Children were informed about the possibility to move their heads while listening to the speech stimuli.

6.2.4 Experimental design

The experimental design relied on the LiSN-S paradigm by Cameron, Dillon, and Newall (2006), with the test conditions conducted in a free-field environment to test target sentence recognition in the presence of distractors with varying SNRs using English speech materials. For the current study, these tests were performed in VAEs with simulated room acoustics, utilising German speech materials.

Speech material

Hochmair-Desoyer et al. (1997) developed a database of German everyday sentences to test SiN perception in individuals fitted with cochlear implants. Out of the 30 lists, a set of 5-word sentences was selected and used as target speech material. The distractors were assigned passages of eight different fairy tales by the Brothers Grimm. The two types of speech materials were recorded by three female native German speakers in the hemi-anechoic chamber, cf. Pausch (2022, Sec. 2.4), presented at a normal speaking pace in a conversational style. For the recordings, a condenser microphone (TLM 170, Georg Neumann GmbH, Berlin, Germany), set to cardioid directivity pattern, was used in combination with a portable audio interface (H6, Zoom Corporation, Tokyo, Japan), operated at a sampling frequency of $f_s = 44.1\,\text{kHz}$.

Table 6.1 presents speech characteristics, analysed for the different female speakers using Praat (Boersma & Weenink, 2016). The results on fundamental frequencies and first and second formants represent average values, obtained from a 60 s long sample section of the same Grimm fairy tale spoken by the different talkers. While the average fundamental frequency, and the first formant of Talker A both were a bit higher than that of Talkers B and C, the latter

Table 6.1: Selected speech characteristics of the three female talkers, presenting the speech materials in Experiment 3. Talker A: Voice of the target talker T and distractors D_1 and D_2; Talkers B and C: Voices of distractors D'_1 and D'_2, cf. Figure 6.2. (Table recreated from Peng, Pausch, and Fels, 2021, and adapted.)

Talker	Fundamental frequency (Hz)	Formant (Hz) 1	2	Speech rate (syllables/s)
A	213	679	1478	3.2
B	191	580	1630	3.2
C	198	562	1611	3.4

two talkers shared similar speech characteristics overall, all talkers exhibiting comparable speech rates.

Test conditions

Figure 6.2 shows the original four test conditions of the LiSN-S paradigm. All sources, i.e. the target talker T and the pairs of distractors $D_{\{1,2\}}$ and $D'_{\{1,2\}}$, are located in the horizontal plane, the target talker being always in the front at 0 deg azimuth. Talker A was always used as target talker and also represented the distractor voices D_1 and D_2, whereas Talkers B and C were used interchangeably only as distractors D'_1 and D'_2, respectively. In the top left panel, the baseline condition is presented, showing the target talker as being co-located with the two distractors D_1 and D_2 (factor *Location* at the level "Co-located", coded as L1), both presenting the distractor speech material with the target talker's voice, i.e. using Talker A (factor *Voice* at the level "Same", coded as V1). In the top right panel, a voice cue was introduced by changing the voices of the distractors D'_1 and D'_2, i.e. using Talkers B and C (factor *Voice* at the level "Different", coded as V2), all sources still being located at 0 deg azimuth. Moving from the baseline condition to the bottom left panel, access to spatial cues was granted by symmetrically shifting the two distractors D_1 and D_2, sharing the target talker's voice, by ±90 deg to the left and right (factor *Location* at the level "Separated", coded as L2). A combination of talker and spatial advantages was provided by additionally replacing the distractor voices using Talkers B and C, while they were likewise separated from the target, as shown in the bottom right panel. Each child underwent testing in these four conditions implemented in the two acoustic versions of the simulated classroom with low reverberation (factor *Reverb* at the level "Low", coded as R1) and high reverberation (factor *Reverb* at the level "High", coded as R2), cf. Figure 6.1, resulting in eight test conditions in total.

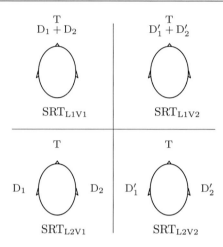

Figure 6.2: LiSN-S test scenarios to measure SRTs with varying distractor locations and voice cues. Top row: Co-located arrangements (coded as L1) of a target talker T and the two distractors D_1 and D_2, sharing the same voice as the target talker (coded as V1), i.e. condition SRT_{L1V1}, or D_1' and D_2', differing in voice (coded as V2), i.e. condition SRT_{S1V2}, all talkers located in the horizontal plane at 0 deg azimuth. Bottom row: Symmetric distractor configurations with D_1 and D_2 (and D_1' and D_2') angularly shifted by ±90 deg azimuth (coded as L2), with the target talker location still fixed at 0 deg azimuth (conditions SRT_{L2V1} and SRT_{L2V2}). Each condition was simulated in low and high RTs, coded as R1 and R2, respectively, cf. Figure 6.1.

Estimation of speech reception thresholds

Estimating SRTs was based on an adaptive up–down staircase procedure (Levitt, 1971), changing the SNR between the target talker and the distractors as per verbal answers provided by the children, in order to converge at the 50 % SRT, cf. Figure 2.12. While the level of the distractors always remained at 55 dB SPL, the target talker started at a level of 70 dB, corresponding to an initial SNR of 15 dB, and was varied in level thereafter. The temporal events per trial per test condition are outlined in the following. After presenting the distractor story for 3 s at the beginning of each trial, a leading beep (1 kHz, 200 ms length) was played back 500 ms before the start of the target sentence, after the end of which the playback of the distractor story was also stopped. Children were instructed to verbally repeat the target sentence. The experimenter, sitting outside the hearing booth, listened to the answers, picked up by the microphone installed inside the booth (C451 E and CK 8, AKG Acoustics, Vienna, Austria), via headphones (HD 650, Sennheiser, Wedemark, Germany), and recorded the number of correctly repeated words of the target sentence. In case of three or more correct words, the SNR was decreased by an initial step size of 4 dB. After the first wrong answer, i.e. the first reversal, the SNR was increased by 2 dB, representing the step size used in all remaining trials in the current condition. Each test condition ended after seven reversals, and SRTs were calculated as arithmetic average of the last four reversals. A maximum playback level of 80 dB SPL in loudspeaker-based binaural playback prevented harmful exposure to the hearing system, particularly in children with NH, and was set as the maximum input level of the MHA of the children with HL.

Procedure

Prior to the start of the experiment, the parents provided written consent for their children. The individual head dimensions were subsequently measured and the child was seated at the chair in the hearing booth. As per LiSN-S instructions, cf. Cameron and Dillon (2007, Appendix C), a monaural audio file was played back via the two front loudspeakers, in which a male speaker introduced the experiment as a game and instructed the children to listen to the target talker in the front after the leading beep, while ignoring any other talkers, and verbally repeat the target sentence. In a training session, all children got presented 3-word sentences to familiarise themselves with the task. The training was completed once 100 % of the first five sentences were correct, or by reaching a cumulative score of 80 % thereafter, which was accomplished by all children. During the actual test blocks, the experimenter started each trial manually and recorded the correctly repeated words contained in the target sentence, before manually triggering the start of the next trial. The talker assignment to distractors D_1' and D_2', and the assignments of fairy tales, were counterbalanced using Latin squares, while the target talker sentences were selected randomly. When necessary, the child was given a break before continuing with the task. At the end of the experiment, the children were rewarded with an allowance of € 10.

6.2.5 Hypotheses

The initially formulated research questions led to four hypotheses, which are presented in Table 6.2.

Table 6.2: Hypotheses of Experiment 3.

Code	Hypothesis
H1-Exp3	Increased reverberation has detrimental effects on SRTs, particularly in children with HL.
H2-Exp3	Children with HL show less benefit in terms of SRTs when the distractors use a different voice than the target voice compared to children with NH, particularly in high reverberation.
H3-Exp3	Children with HL show less SRM compared to children with NH, particularly in high reverberation.
H4-Exp3	The SRT values achieved by children fitted with research HAs differ from those achieved by children with HL using their own commercial HAs.

6.3 Results

The data analysis described below was conducted using R (R Core Team, 2021) and RStudio (RStudio Team, 2021). All results were interpreted at a confidence level of 95 %.

6.3.1 Missing data

Proportions of both participant groups were not able to complete all experimental conditions owing to various reasons, such as, for example, headache, lack of concentration, and fatigue. In total, 11 children with HL and ten children with NH aborted the experiment, resulting in six and 18 complete data sets, respectively. Figure A.2 visualises the pattern of missing data per participant and experimental conditions, split into groups. The final dataset exhibited mean percentages of missingness of $32.4 \pm 15.4\,\%$ ($\mu \pm \sigma$, range: 11.8–52.9 %) and $16.1 \pm 13.5\,\%$ ($\mu \pm \sigma$, range: 3.6–32.1 %) by experimental condition and groups of children with HL and children with NH, respectively, containing a total of 22.2 % of missing data. Non-significant Little's test results (Tierney et al., 2021), $p > .05$, performed by group, indicated that the data in both data sets are missing completely at random.

6.3.2 Speech reception thresholds

Table 6.3 and Figure 6.3 present the estimated SRTs by experimental condition and participant group. The results from participants having worn their own HAs are represented by the grey error bars. Data analysis followed the procedure described below for testing the hypotheses presented in Table 6.2.

Table 6.3: Estimated SRTs by experimental condition, split into participant groups. Additional comparison values are provided for children who used their own HAs.

Group	Talker	Location	Reverb	SRT (dB) Core sample $\mu \pm \sigma$	Children w/ own HAs $\mu \pm \sigma$
HL	Same	Co-located	Low	11.8 ± 6.3	12.6 ± 3.7
			High	13.4 ± 6.1	15.9 ± 4.0
		Separated	Low	9 ± 4	10.4 ± 6.7
			High	11 ± 4.7	12 ± 3.9
	Different	Co-located	Low	10.9 ± 6.2	11.4 ± 5.0
			High	11.3 ± 6.2	14.4 ± 3.3
		Separated	Low	7.6 ± 5.1	12 ± 5.2
			High	11.7 ± 7.2	5.9 ± 6.0
NH	Same	Co-located	Low	4.4 ± 3.4	
			High	7.1 ± 2.2	
		Separated	Low	-2.2 ± 4.6	
			High	0.7 ± 3.7	
	Different	Co-located	Low	3.6 ± 4.2	
			High	5.3 ± 2.5	
		Separated	Low	-3.2 ± 5.5	
			High	0.6 ± 4.9	

To avoid listwise deletion owing to missing data, an LME model was formulated and fit by restricted maximum likelihood estimation (Bates et al., 2015) with unconstrained and bounds-constrained quasi-Newton method optimiser (Nash & Varadhan, 2011; Nash, 2014) to predict the SRTs in the core sample (249 data points in total). As fixed effects, I entered the factors *Group* at the levels HL and NH, representing data associated with the children with HL and children with NH, respectively. Additionally, *Voice*, at the levels Same and Different, *Location*, at the levels Co-located and Separated, as well as *Reverb*, at the levels Low and High, were also included as fixed factors. Interaction terms between all fixed factors were part of the model as well. The random effects structure included four random-intercept terms, one for participants, and three for participants nested within *Voice*, *Location*, and *Reverb*. Diagnostic plots were evaluated and indicated no violations of model assumptions. Applying the same backward elimination procedure as in Experiment 1, see Chapter 5, Section 5.1.2, the results on random effects suggested to remove the nesting of participants within *Reverb* and *Talker*, $p > .05$ (Halekoh & Højsgaard, 2014). Regarding the fixed effects, all interaction terms except *Group × Location* were identified as being redundant,

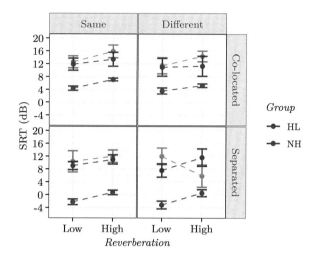

Figure 6.3: Estimated SRTs by experimental conditions, split into participant groups. Dots and error bars indicate means and standard errors of the mean, respectively. Grey dots and error bars refer to the results of children with HL who used their own HAs.

$p > .05$. Again, diagnostic plots on the reduced model were free of conspicuities with regard to violations of model assumptions.

Likelihood ratio tests after single term deletions (function drop1 from R package stats), were performed on the reduced model[2]. The test results indicated main effects of *Voice*, $F(1, 175.31) = 6.54$, $p = .011$, and *Reverb*, $F(1, 170.42) = 61.03$, $p < .001$. Post-hoc analysis revealed significantly elevated mean SRTs when the distractors shared the same voice as the target talker ($\mu \pm \sigma = 4.8 \pm 6.4$ dB, range: -10.1 to 22.4 dB), compared to conditions in which the distractor voices were different from that of the target talker ($\mu \pm \sigma = 3.4 \pm 6.6$ dB, range: -10.7 to 21.6 dB). Manipulating the RT also entailed increased SRTs in high reverberation ($\mu \pm \sigma = 5.5 \pm 5.9$ dB, range: -5.6 to 21.6 dB) versus low reverberation ($\mu \pm \sigma = 3 \pm 6.8$ dB, range: -10.7 to 22.4 dB).

The interaction *Group* × *Location* was significant, $F(1, 37.43) = 7.74$, $p = .008$, indicating a group-dependent change of SRTs between conditions with co-located target talker and distractors (HL: $\mu \pm \sigma = 12 \pm 5.9$ dB; NH: $\mu \pm \sigma = 5.2 \pm 3.3$ dB) versus conditions with spatially separated distractors (HL: $\mu \pm \sigma =$

---

[2]Note that these tests were conducted following the principle of marginality, not displaying main effects when involved in higher-order interactions (Venables, 1998).

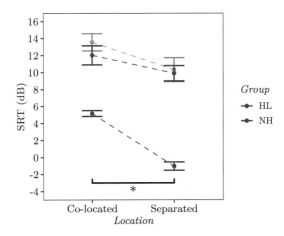

Figure 6.4: Significant interaction effect between *Group* × *Location*. Dots and error bars indicate means and standard errors of the mean, respectively. Grey dots and error bars present the results of children with HL who used their own HAs.

$9.9 \pm 5.2\,\text{dB}$; NH: $\mu \pm \sigma = -1 \pm 4.9\,\text{dB}$), cf. Figure 6.4. Analysis of simple effects using multiple t-tests with Holm-Bonferroni correction indicated a non-significant $2.3\,\text{dB}$ SRM in children with HL, $t(39.8) = 1.9$, $p = .065$, but a significant $6.2\,\text{dB}$ SRM in children with NH, $t(31.9) = 8.6$, $p < .001$. Between groups, significant SRT differences of $7.5\,\text{dB}$ in conditions with co-located target talker and distractors, $t(57.7) = 5.13$, $p < .001$, and $11.4\,\text{dB}$ after spatially separating the distractors, $t(53.7) = 8.03$, $p < .001$, were present.

The core analysis revealed no other main effects or interactions in SRT data, $p > .05$.

The attempt to investigate Hypothesis H3-3 based on a similar LME model as formulated for the core analysis, with a between-group factor consisting of the children with HL, split into data cells of children wearing the research HAs and children using their own HAs, resulted in rank-deficient fixed-effect model matrices and non-converging results. Particularly for reasons of small sample size in the group of children using their own HAs and consequent normality issues, I decided to conduct eight unpaired two-samples Wilcoxon signed-rank tests between the associated data cells, see first eight rows in Table 6.3. All results suggested non-significant differences of population mean ranks in SRT values, $p > .05$, in all test conditions between the two sub-groups of children with HL aided differently.

6.4 Discussion

6.4.1 Main effects

Overall effect of reverberation

Numerous studies investigated the impact of RT on speech intelligibility in children with HL and children with NH (see, e.g., Finitzo-Hieber & Tillman, 1978; Neuman & Hochberg, 1983; Klatte, Lachmann, & Meis, 2010; Iglehart, 2020) and came to the common conclusion that excessive RTs have a substantial negative effect on speech perception. Iglehart (2020) tested children fitted with HAs of a similar age ($\mu = 11.4$ yrs, range: 7.1–16 yrs) to the current sample in varying classroom RTs of 0.3 s, 0.6 s, and 0.9 s for 50 % word recognition utilising the Bamford–Kowal–Bench SiN test. Since the scores improved significantly after each reduction of RT, the author supported the current recommendations by ANSI/ASA S12.60-2010/Part 1 (2010) on RTs of 0.3 s in learning spaces with volumes smaller than 283 m^3. The findings of the current study once again corroborate worse SiN perception, quantified by significantly increased accumulated SRTs in both groups of children. Compared to the data reported in Cameron and Dillon (2007, Table 3) measured in free-field conditions, the SRT results obtained in the current study were consistently higher, a deviation that has been further increased particularly by prolonging the RTs. This discrepancy supports the conclusion by Iglehart (2020) that clinical assessments in free-field conditions overestimate the SiN perception performance in children with HL fitted with HAs. Since there were no significant interactions in the core analysis, increasing the RTs seemed to have deleteriously affected both groups of children, regardless of the introduced voice and spatial cues, thus only supporting Hypothesis H1-Exp3 in terms of an overall negative effect on SiN perception performance.

Overall effect of the voice cue

Based on the results in the current study, both groups of children seemed to benefit equally from the introduced voice cue, entailing significantly lower SRTs when the distractors had different voices than the target talker. However, given an accumulated voice advantage of only 1.4 dB, the overall effect seems to be not very pronounced, and also less salient compared to the accumulated differences obtained under free-field conditions (Cameron & Dillon, 2007, Table 4). Reasons for a reduced effect can be in part likely traced back to temporal effects of reflections, hindering to make use of spectral and temporal glimpsing by smearing the spectrum of speech and flattening its temporal envelope, respectively (Houtgast & Steeneken, 1985). A benefit after separating the distractors from the target talker and thus increasing access to binaural cues as well as auditory grouping mechanisms (Bregman, 1994) could not be detected in the current sample of children. Also, decreasing RTs did not help to improve exploitation of diverging speech characteristics. Collectively, these results only support Hypothesis H2-Exp3 regarding an overall voice advantage, without specifically benefiting any of the investigation groups, neither in scenarios with co-located target talker and distractors or when arranged as spatially separated, nor in low or high RTs.

6.4.2 Spatial release from masking

Most interestingly, the significant two-way interaction revealed a group effect in SRM, indicating that children with HL overall could not make use of spatial cues as efficiently as their peers with NH, with signficant group differences as observed for SiN perception with energetic maskers (Nittrouer et al., 2013; Torkildsen et al., 2019). Marrone, Mason, and Kidd (2008) tested SiN perception in young adults with NH ($\mu = 31.4$ yrs) and young adults with HL ($\mu = 28.5$ yrs). The experiments were performed in a sound booth in two acoustic conditions with RTs of 0.06 s (condition BARE) and 0.25 s (condition PLEX) by physically replacing the absorbers with perspex panels. Stimuli were presented via three discrete loudspeakers from 0 deg and ±90 deg azimuth. Although the acoustic conditions are only roughly comparable to the ones simulated in the current study, substantially lower SRTs were measured in both groups throughout test conditions with co-located target and distractors compared to the current results. Their reported significant interactions between RT and spatial separation, as well as between RT and group, were not found in the current investigation with children. But their group effect on SRM with similar pattern was corroborated by the current two-way interaction *Group* $\times$ *Location*. However, compared to their PLEX room condition, lower overall SRM values of current 6.2 dB (significant) versus about 7.9 dB (significant) in young adults with NH, and current 2.3 dB (not significant) versus about 3 dB (significant) in young adults with HL were measured. Similarly as reflected by increased SRTs (Iglehart, 2020), the trend towards decreased SRM may be linked to the influence of higher RTs (Marrone, Mason, & Kidd, 2008) and their aforementioned effects on binaural cues and auditory grouping abilities. Further specific investigations in scenarios exhibiting RTs in the range of target RTs of 0.3 s for classrooms $\leq 283\,\mathrm{m}^3$ (ANSI/ASA S12.60-2010/Part 1, 2010) and 0.48 s for the current classroom with $244\,\mathrm{m}^3$ (DIN 18041, 2016) are required to re-evaluate whether the recommended limits for inclusive education are still appropriate for children with HL. Overall, Hypothesis H3-Exp3 is partially supported by the current results, lacking to corroborate an effect when varying RTs and voice characteristics of the distractors.

6.4.3 Validity of results

Statistic test results suggested that the sample of children with HL fitted with research HAs obtained similar SRTs in all test conditions compared to the results of children with HL using their own HAs. Although given low statistical power due to the small sample size and missing data, these results attribute a certain degree of reliability to the system to have the potential of achieving valid results when using the HA simulation as integral part of the system, which goes against Hypothesis H4-Exp3. The system performance determined perceptually in this way can thus be regarded as a reference for variations in future experiments on SiN perception involving listeners with HL, while being immersed in simulated room acoustics. Such variations may include the integration of HA algorithms (e.g. Healy et al., 2013), and more refined modifications in room acoustic simulations (e.g. gradually increasing RTs, manipulating C_{50} values, or changing IACC), or the characteristics (energetic vs. informational maskers), number, distance and movement of maskers (for a review, see Viveros Munoz, 2019, Ch. 3).

6.4.4 Limitations

Although individual measurements of HRTFs and HARTFs could have further increased the validity of the results, the presented individualisation approach was preferred, mainly because of the otherwise increased effort for teachers, parents and children. For the same reasons, no individual real-ear-to-coupler differences were measured, potentially having led to biased real-ear gains that differ from the prescriptive target gain. A more playful design with virtual interim rewards after each test condition may have sustained the children's endurance and possibly led to less terminations. To shorten the total duration of the experiment given the already high number of test conditions, maximum-likelihood procedures to estimate SRTs would favour to reduce the number of required trials per test block until convergence, cf. Section 2.9.

6.5 Conclusion

An established binaural SRM paradigm was implemented in virtual classroom environments with low and high RTs, which were rendered and reproduced applying a binaural real-time auralisation system. Children with HL, bilaterally fitted with research HAs, and children with NH were tested for SiN perception, utilising a sentence recognition task. The results of a small group of children with HL using their own commercial HAs were used to check the validity of the applied HA simulation. Returning to the initially posed research questions, the major findings are summarised below:

1. Measured SRTs equally decreased when lowering the RTs from 1.2 s to 0.4 s, aggregated across the remaining test conditions and investigation groups. Generally higher SRTs were measured throughout test conditions in both investigation groups compared to results obtained in free-field environments using the same paradigm.

2. The introduced voice cue benefited both groups to a similar amount but revealed no intertwined effect with the location of the distractors relative to the target talker, being also unaffected by the amount of reverberation.

3. No effect of RT, but a group effect on SRM was found. Specifically, children with NH were able to benefit from SRM, accumulated across the remaining test conditions including the two RTs and distractor voice conditions, while children with HL only showed tendencies to improve their performance provided spatial separation of the distractors. In the light of these results, it was concluded that children with HL fitted with research HAs in the tested age range were unable to cope with the disruptive nature of room reflections with RTs longer than 0.4 s.

4. Comparable results in terms of SRTs, measured in children wearing their own commercial HAs and children fitted with research HAs, indicated that using the proposed approach of integrating HAs in the virtual scene allows to accurately quantify SiN perception.

7

Conclusion

7.1 Summary

The main objective of this thesis was to elaborate novel concepts that allow for transparently integrating research HAs in state-of-the-art VAEs, validate their specific implementations via objective and perceptual evaluations, and demonstrate applicability for clinical purposes.

In Chapter 3, such a concept was developed considering the combined perception of sound fields by aided listeners with HL via HAs and utilising residual hearing. The concept also addresses general requirements regarding complex acoustic scene generation of physically based indoor environments involving multiple static or moving VSSs, transparent signal rendering throughout all stages, real-time processing and reproduction of HA signals, external sound field reproduction, and the reproduction environment. In order to approach the overarching goal of ecological validity, the system design was extended to include real-time implementations and room acoustic simulations, allowing to account for listener movements via motion tracking and simulate plausible indoor environments, respectively. Possible loudspeaker-based spatial audio reproduction methods, either based on binaural technology or panning-based methods, were introduced for external sound field simulation. The careful fusion of design criteria resulted in two specific system implementations of HAA systems:

1. Underlying the conditions of a complex scene, the combined binaural real-time auralisation system relies on simulated binaural HA signals, which are based on measured HARTFs, representing the signals at the HA microphone positions. After processing these signals on an MHA, including and parametrising HA algorithms as required, playback is accomplished via the HA receivers. External binaural sound field reproduction was implemented through an array of loudspeakers with N'-CTC filters, rendering the au-

ralisation approach as entirely binaural. It was made possible to adjust the relative delay between the two reproduction paths that would occur in real-life scenarios due to HA algorithm processing delays via a VDL. An optical motion tracking system facilitates capturing listener movements to continuously update the rendering and reproduction stages by application of efficient real-time convolution algorithms and room acoustic simulation models. Both real-time auralisation and simulation of room acoustics access pre-generated generic databases of rendering HARTFs and HRTFs with high spatial resolution. Possible simulation configuration scenarios were defined to be evaluated for their suitability given limited processing power on desktop computers.

2. A hybrid static approach was additionally introduced enabling complementary systematic objective and perceptual investigations. Depending on the selected spatial audio reproduction method, the choice of which may be N'-CTC, VBAP, MDAP or HOA, the corresponding loudspeaker signals are rendered, optionally including room acoustic simulations. To study the influence of the reproduction environment on measurement results and perception, the reproduction signal flow was extended to include in-situ playback transfer functions between the loudspeakers and the receiver at the array centre. The subsequent convolution with the loudspeaker signals affects the simulated HA signals, the ear signals utilising residual hearing, as well as their summed combination.

Chapter 4 globally analysed the acoustic characteristics and differences between reference rendering HRTF and HARTF databases with high spatial resolution in terms of ITDs and ILDs, as well as monaural cues. A benchmark analysis of an example scene served to assess the aforementioned simulation configuration scenarios. When the HAA system is operated on a desktop computer with limited processing resources, it was concluded that a configuration with pre-calculated reflections is best suited for complex acoustic scenes involving room acoustic simulations with long filters and multiple VSSs (Table 3.1, Configuration A). If the number of VSSs is low and compromises can be made regarding the image source order, in combination with restricting simulation and filter update rates as per perceptual demands, Configurations B and C can be considered as well. However, full real-time room acoustic simulation and filter updates as per Configuration D is reserved to cluster-based computing. The measurement of the combined static system EEL using Configuration A and 6-channel filter lengths of 44,100 samples each amounted to 25.9 ms and confirmed the pre-defined relative delay between reproduction paths. Combined dynamic EEL was estimated to be about 80 ms based on results from latency measurements using a comparable motion tracking hardware, which only slightly exceeds the recommended 50–75 ms (Brungart, Simpson, & Kordik, 2005; Yairi, Iwaya, & Suzuki, 2006; Vorländer, 2020). For further objective evaluation, spatial in-situ playback transfer functions were measured in the commercial and custom-made hearing booths, the VR laboratory, and the anechoic chamber. The measurement results largely show high consistency with the corresponding ideal playback transfer functions, measured in the hemi-anechoic chamber, apart from high frequency deviations emerging from detrimental reflections whose influence is subject to evaluation. Investigations on the effect of regularisation in binaural loudspeaker-based reproduction suggested that setting β to 0.01 represents a good compromise in terms of CS and perceptual quality. Substantial differences between optimal and achievable CS corroborated the detrimental effect of reflections, as previously reported in the literature. Consistently higher broadband achievable CS results were obtained in the custom-made hearing booth compared to the commercial one, hinting increased effectiveness of the acoustic optimisations in the former reproduction environment. It was shown that increasing the number of involved loudspeakers N' in the two high-channel loudspeaker arrays had a positive effect on both optimal and achievable CS, particularly in the high frequency range. Mean increases of about $2 \pm 1.8\,\mathrm{dB}$ ($\mu \pm \sigma$) and $3.2 \pm 0.8\,\mathrm{dB}$ ($\mu \pm \sigma$) per loudspeaker doubling were observed for the setups in the VR laboratory and the anechoic chamber, respectively, suggesting to use a number of $N'_{\mathrm{VR}} = 16$ and $N'_{\mathrm{AC}} \geq 8$ for sufficient CS. The evaluation of SDs in recreated receiver directivities, i.e. HRTFs, and front and rear HARTFs, for directions on a spatial grid revealed a larger influence of reflections in the VR laboratory above 1 kHz, which was less pronounced for the setup in the anechoic chamber. Panning-based methods additionally entailed a consistent error variance for frequencies below. Large effects of room reflections were observed for the ITDs in recreated receiver directivities using panning-based loudspeaker reproduction, in particular in the VR laboratory, with consistently lower errors using loudspeaker-based binaural reproduction. The ILD errors increased especially for binaural loudspeaker reproduction in the presence of reflections, while all panning-based methods showed generally increased errors with similar error tendencies and variances, which were largely unaffected by the used in-situ spatial playback transfer function type. Since special attention was paid to the influence of the reproduction environment, the transferability of the current results to similar non-ideal laboratories is favoured.

In Chapter 5, the two specific system implementations were perceptually evaluated by investigating two essential spatial audio quality parameters to determine the egocentric position of a sound source. Experiment 1 compared the dynamic localisation performance of RSSs with that observed for VSSs, binaurally reproduced via headphones, loudspeakers with CTC filters, as combined via loudspeakers with CTC filters and research HAs, and via research HAs alone, each time simulated in a free-field environment. Pooled reversal rates gradually increased in order of the listed individual systems, with highest rates of about 28 % given playback via research HAs alone, decreasing to about 19 % in combined binaural playback with loudspeakers and CTC filters. A positive and significant effect of binaural loudspeaker playback was observed in combined reproduction regarding overall horizontal source localisation performance, corroborating the trend in reversal rates, which, however, was not present in the evaluated angular error measures. The fact that both azimuth and elevation errors in binaural loudspeaker playback and combined playback were comparable allowed to conclude that the deterioration in localisation performance can be primarily attributed to largely inaccessible ITDs in the low frequency range, rather than to the influence of distorted monaural cues in HA-based playback.

The effect of the reproduction method on egocentric auditory distance estimation in simulated room acoustics, following the concept of the combined static hybrid auralisation approach using individually measured high-resolution rendering and in-situ playback receiver directivities, was investigated in Experiment 2. The resulting coefficients after linearly fitting a compressive power function to the distance estimates obtained in binaural headphone reproduction showed high similarity to previously reported average results in the literature, given binaural headphone playback based on measured BRIRs. Significantly lower ordinate intercepts were present in the remaining conditions, when presenting the VSSs via research HAs, loudspeakers with CTC filters, VBAP, HOA, and as combined via the loudspeaker-based methods and the research HAs. Although the slopes ζ tended to increase compared to headphone-based reproduction, only the contribution of HOA had a significant effect on ζ when testing the influence of loudspeaker-based reproduction in combined playback. A follow-up analysis revealed lower DRRs in the simulated signals, which were considered the main reason for the substantially higher auditory horizons in the current study throughout all conditions compared to previous results. No effect of the reproduction method on the RMS errors in distance estimates was observed. It was concluded that egocentric auditory distance estimation in participants with NH via research HAs alone and in combined reproduction is comparable to the performance when presenting VSSs via the individual loudspeaker-based methods.

Collectively, the perceptual results indicate that the acoustic discrepancies between HARTFs and HRTFs, as well as the HA receiver characteristics have more pronounced effects on localisation performance than egocentric auditory distance perception.

Chapter 6 focused on the application of the combined real-time auralisation system in a clinical experiment, testing the SiN perception performance in school-age children with HL, fitted with research HAs, and age-matched children with NH in simulated classroom environments given low and high RTs. It was shown that the SRTs were generally higher as the ones obtained in free-field environments using the same paradigm. A significant decrease of SRTs was found in both groups when lowering the RTs, further motivating the importance of optimised room acoustics in critical learning spaces. Both groups showed equal amount of benefit when the voice of the distractors were different from that of the target talker, which, however, was not influenced by the distractors' locations and the simulated RTs. Most interestingly, spatially separating the distractors from the target talker was only of use for children with NH. Since children with HL were not able to benefit, it was concluded that RTs above 0.4 s already entail a critical disruption of spatial processing in this vulnerable group. Comparable SRTs were obtained in children fitted with research HAs compared to those wearing their own commercial HAs, demonstrating the potential of the HA simulation.

7.2 Outlook

The work presented was only able to cover a small portion of evaluation and application possibilities of the proposed systems and also revealed existing shortcomings. Some ideas for improvements and future directions are therefore provided below.

Although efforts have already been made to improve CS in loudspeaker-based binaural reproduction systems in the presence of reflections (Sæbø, 2001; Kohnen et al., 2016), the effectiveness of these measures is not yet satisfying and continues to be a challenge for current research. As already hinted by Sanches Masiero (2012), approaches involving dereverberation of room IRs (Jungmann et al., 2012) may be improved by exploiting psychoacoustic effects, such as temporal masking thresholds, during CTC filter design. To reduce the occurrence of reversal rates in binaural reproduction systems, particularly when using generic datasets of receiver directivities, the application of directional weighting functions may also be beneficial for CTC playback and sound field perception via HAs (Blauert, 1997; Møller et al., 1995a; Lee, Kim, & Sung, 2013).

Since desktop computers only provide limited processing power, compromises have to be made with regard to room acoustic simulations in terms of filter lengths, filter update rates, and the number of possible VSSs. Once again, perceptually motivated techniques like culling (Blauert, Mourjopoulos, & Buchholz, 2001; Begault, McClain, & Anderson, 2004) or clustering (Moeck et al., 2007) likely allow to expanding such frontiers and help to further reduce hardware costs, which overall may be a constraining factor for financially disadvantaged institutions to use such auralisation systems. In combination with measurement-based knowledge about the latency caused by the motion tracking system, it is possible to estimate whether the dynamic EEL is within perceptual limits in increasingly complex acoustic scenarios after system optimisations. More investigations are necessary to specify perceptually required spatial resolutions of databases (Lindau, Maempel, & Weinzierl, 2008) and filter update rates (Lentz, 2008). Even though the systems are already very powerful, the real-time calculations and scene interactivity should generally be reduced to the necessary minimum in auditory experiments to maintain reproducibility and controllability.

The system evaluations remain to be extended to include specific investigations of HA algorithms and their perceptual influence, which was out of scope in the current thesis but already initiated by Vollmer (2018). Following similar lines of reasoning as Cubick and Dau (2016), discrepancies between a simulated environment and its real-world equivalent need to be identified by systematic modifications to improve the simulation and consequently increase the prediction quality on behavioural outcome measures. Specifically designed acoustic indoor scenarios in simulated room acoustics to be used as training data for deep neural networks may take the performance of HA algorithms to the next level (Vivek, Vidhya, & Madhanmohan, 2020).

In the context of system application, the proposed systems can also be applied for modelling purposes (Viveros Munoz, 2019). Not only the behavioural results from participants with HL, but also from other investigation groups like, for example, individuals with attention deficit hyperactivity disorder, or a suspected central auditory processing disorder, represent important research data. Especially to support children in the mentioned groups to better cope with challenging learning environments as they grow up, the possibility to plausibly simulate complex acoustic scenes that resemble those of everyday life scenarios opens up a multitude of possibilities in the context of auditory training (Cameron, Glyde, & Dillon, 2012; Henshaw & Ferguson, 2013).

Finally, the modular framework of the system does not limit its implementation to the components presented, but should motivate other research groups to integrate similar ones that are readily available as open-source software (for a review, see Pausch et al., 2018b) and hardware used in their laboratories. As an example, the system design would also allow expanding research to cochlear implant users by replacing the MHA by dedicated research platforms (Shekar & Hansen, 2021). Just dealing with the system encourages new ideas for improvement and its use in scientific studies as a flexible tool. Hopefully, the research results will benefit both individuals with HL and those with NH to improve the quality of life.

Appendix

Supplementary material

The appendix contains supplementary material for Experiment 1, cf. Chapter 5, and Experiment 3, cf. Chapter 6.

A.1 Experiment 1

Table A.1 presents the formulated LME models in tabular form (Lüdecke, 2021). Model 1 and Model 2 have identical structure and complexity but refer to condition LS and condition RHA, respectively. The mean coefficient estimates and their 95 % CIs are presented with Holm-Bonferroni-corrected p-values for the fixed effects, obtained from t-tests applying Kenward-Roger's approximation for estimating the degrees of freedom in the denominator. Significant results are written in bold letters. The random effects are complemented by σ^2 and τ_{00}, with indices indicating the nesting of participants, and adjusted and conditional intra-class correlation (ICC) coefficients. Marginal and conditional R^2 values, as well as AIC values provide further information about the model fit.

Table A.1: Coefficient summary of the LME model, applied for fitting the horizontal localisation results in Experiment 1 by restricted maximum likelihood.

	Perceived azimuth					
	Model 1 (re condition LS)			Model 2 (re condition RHA)		
Coefficients	Estimate	95 % CI	p-value	Estimate	95 % CI	p-value
Fixed effects						
(Intercept)	13.42	[3.32, 23.51]	.066	40.24	[30.15, 50.34]	< .001
HP vs. LS	3.27	[−9.47, 16.01]	1.000			
CTC vs. LS	−4.39	[−17.13, 8.35]	1.000			
RHA vs. LS	26.82	[14.08, 39.56]	< .001			
CTCwRHA vs. LS	5.27	[−7.47, 18.01]	1.000			
Presented azimuth	0.94	[0.88, 0.99]	< .001	0.77	[0.77, 0.82]	< .001
HP vs. LS × Presented azimuth	−0.01	[−0.08, 0.06]	1.000			
CTC vs. LS × Presented azimuth	0.01	[−0.06, 0.08]	1.000			
RHA vs. LS × Presented azimuth	−0.17	[−0.23, −0.1]	< .001			
CTCwRHA vs. LS × Presented azimuth	−0.05	[−0.12, 0.02]	0.774			
LS vs. RHA				−26.82	[−39.56, −14.08]	.001
HP vs. RHA				−23.55	[−36.29, −10.81]	< .001
CTC vs. RHA				−31.22	[−43.96, −18.48]	.002
CTCwRHA vs. RHA				−21.55	[−34.29, −8.81]	< .001
LS vs. RHA × Presented azimuth				0.17	[0.1, 0.23]	< .001
HP vs. RHA × Presented azimuth				0.16	[0.09, 0.22]	< .001
CTC vs. RHA × Presented azimuth				0.18	[0.11, 0.25]	< .001
CTCwRHA vs. RHA × Presented azimuth				0.11	[0.05, 0.18]	.002
Random effects						
σ^2	760.56			760.56		
$\tau_{00,\ Presented\ azimuth:Participant}$	24.16			24.16		
Adjusted ICC/conditional ICC	0.2/0.02			0.2/0.02		
Model fit						
Marginal R^2/conditional R^2	0.899/0.92			0.899/0.92		
AIC	5803.98			5803.98		

A.2 Experiment 3

The results with respect to the individual pure-tone audiograms of the children with HL, and supplementary demographic information are presented in Figure A.1 and Table A.2, respectively.

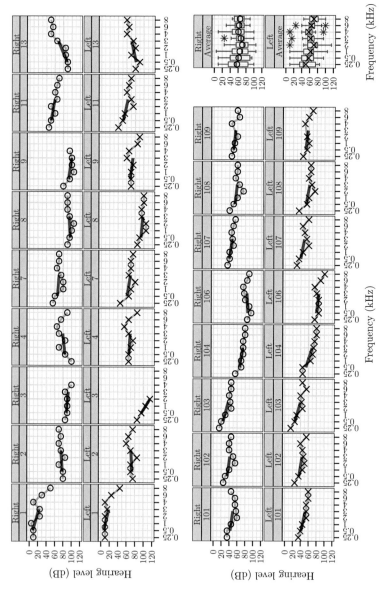

Figure A.1: Individual air-conduction pure-tone audiograms of the children with HL. Solid black lines represent linear regression lines, predicting the hearing thresholds evaluated across frequencies of 0.5, 1, 2 and 3 kHz. Missing data points indicate no response from the child at the respective test frequency. Bottom right: Average hearing thresholds across children with HL, visualised via boxplots with medians and means as horizontal lines and crosses, respectively. Whiskers indicate 1.5 times the interquartile range, and outliers are shown as asterisks.

Table A.2: Demographic information about the children with HL. Participants with IDs marked with asterisks wore their own HAs and were not part of the core analysis. Pooled results with uncertainties represent $\mu \pm \sigma$. Missing information is indicated by n/a.

ID	Gender	Age (yrs)	HA use		Average hearing threshold (0.5–3 kHz, dB)		Slope of regression line (0.5–3 kHz, dB/octave)	
			Days w/ school (h)	Days w/o school (h)	Left	Right	Left	Right
1	m	7.9	12	10–11	11.3	16.3	-1.5	-6.5
2*	f	9.3	12.5	12.5	73.6	77.5	-1.5	0
3	f	13.2	15–16	15–16	105	88.8	-10	1.5
4*	f	8.1	12	12	68.8	77.5	-0.5	3
7	f	10.5	14–15	14–15	70	73.8	-1	-4.5
8	m	11.3	14–15	14–15	105	88.8	-10	1.5
9	m	12.1	13	15	78.8	81.3	2.5	-1
11*	f	8	12	14	62.5	56.3	-8	-3.5
13*	m	12.1	13	15	78.8	81.3	3.5	7.5
101	f	12.8	15	12	48.8	49.4	-5.5	-8.3
102	f	10.9	12.5	11	44.4	43.8	-4.3	-7.5
103	f	10.6	10	8	35.6	34.4	-6.8	-8.3
104	m	10.2	14	14	63.1	72.5	-8.8	-2
106	m	12.7	13	13	83.8	85.6	1.5	6.3
107	m	7.4	12	12	48.8	44.4	-7.5	-3.3
108	f	7.2	11	11	64.4	62.5	-5.3	-6
109	f	10.6	n/a	n/a	55.7	54.2	1.4	-5
Pooled	10f / 7m	10.3 ± 2	12.9 ± 1.5	12.8 ± 2	63.7 ± 22.6	65.7 ± 23.3	-3.2 ± 4.1	-2.2 ± 4.8

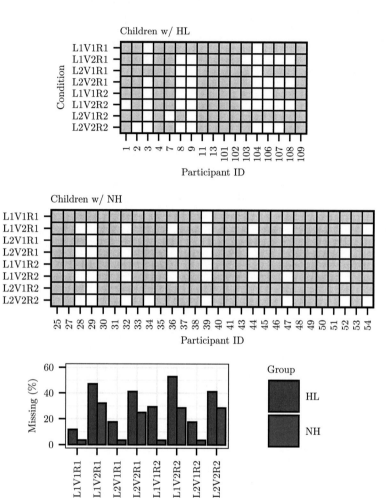

Figure A.2: Visualisation of missing data points as white squares per participant and experimental condition, split into groups of children with HL and children with NH. The lowest panel shows the proportions of missing data by experimental condition and group.

Condition coding:
L1/L2... Co-located/spatially separated target talker and distractors
V1/V2... Target talker and distractors use the same voice/different voices
R1/R2... Low/high RT

List of publications

Peer-reviewed journal articles

Bell, L., Peng, Z. E., Pausch, F., Reindl, V., Neuschaefer-Rube, C., Fels, J., & Konrad, K. (2020). fNIRS Assessment of Speech Comprehension in Children with Normal Hearing and Children with Hearing Aids in Virtual Acoustic Environments: Pilot Data and Practical Recommendations. *Children*, *7*(11). https://doi.org/10.3390/children7110219

MacCutcheon, D., Hurtig, A., Pausch, F., Hygge, S., Fels, J., & Ljung, R. (2019a). Second language vocabulary level is related to benefits for second language listening comprehension under lower reverberation time conditions. *Journal of Cognitive Psychology*, *31*(2), 175–185. https://doi.org/10.1080/20445911.2019.1575387

MacCutcheon, D., Pausch, F., Fels, J., & Ljung, R. (2018). The effect of language, spatial factors, masker type and memory span on speech-in-noise thresholds in sequential bilingual children. *Scandinavian Journal of Psychology*, *59*(6), 567–577. https://doi.org/10.1111/sjop.12466

MacCutcheon, D., Pausch, F., Füllgrabe, C., Eccles, R., van der Linde, J., Panebianco, C., Fels, J., & Ljung, R. (2019b). The Contribution of Individual Differences in Memory Span and Language Ability to Spatial Release From Masking in Young Children. *Journal of Speech, Language, and Hearing Research*, *62*(10), 3741–3751. https://doi.org/10.1044/2019_JSLHR-S-19-0012

Marsh, J. E., Ljung, R., Jahncke, H., MacCutcheon, D., Pausch, F., Ball, L. J., & Vachon, F. (2018). Why are background telephone conversations distracting? *Journal of Experimental Psychology: Applied*, *24*(2), 222–235. https://doi.org/10.1037/xap0000170

Pausch, F., Aspöck, L., Vorländer, M., & Fels, J. (2018b). An Extended Binaural Real-Time Auralization System With an Interface to Research Hearing Aids for Experiments on Subjects With Hearing Loss. *Trends in Hearing*, *22*, 1–32. https://doi.org/10.1177/2331216518800871

Pausch, F., Behler, G., & Fels, J. (2020). SCaLAr – A surrounding spherical cap loud-speaker array for flexible generation and evaluation of virtual acoustic environments. *Acta Acust.*, *4*(5), 19. https://doi.org/10.1051/aacus/2020014

Pausch, F., Doma, S., & Fels, J. (2021). Hybrid multi-harmonic interaural time difference model for individual behind-the-ear hearing aid-related transfer functions [Unpublished manuscript, in review]. *Institute for Hearing Technology and Acoustics, RWTH Aachen University.*

Pausch, F., & Fels, J. (2019b). MobiLab – A Mobile Laboratory for On-Site Listening Experiments in Virtual Acoustic Environments. *Acta Acustica united with Acustica*, *105*(5), 875–887. https://doi.org/doi:10.3813/AAA.919367

Pausch, F., & Fels, J. (2020). Localization Performance in a Binaural Real-Time Auralization System Extended to Research Hearing Aids [PMID: 32324491]. *Trends in Hearing*, *24*, 1–18. https://doi.org/10.1177/2331216520908704

Peng, Z. E., Pausch, F., & Fels, J. (2021). Spatial release from masking in reverberation for school-age children. *The Journal of the Acoustical Society of America*, *150*(5), 3263–3274. https://doi.org/10.1121/10.0006752

Pomberger, H., & Pausch, F. (2014). Design and Evaluation of a Spherical Segment Array with Double Cone. *Acta Acustica united with Acustica*, *100*(5), 921–927. https://doi.org/10.3813/AAA.918771

Technical reports

Pausch, F. (2022). *Documentation of the experimental environments and hardware used in the dissertation "Spatial audio reproduction for hearing aid research: System design, evaluation and application"* (tech. rep.). Institute for Hearing Technology and Acoustics, RWTH Aachen University, Germany. https://doi.org/10.18154/RWTH-2022-01536

Preprints

Pausch, F., & Fels, J. (2019a). MobiLab – A Mobile Laboratory for On-Site Listening Experiments in Virtual Acoustic Environments. *bioRxiv*. https://doi.org/10.1101/625962

Contributions to international conferences

Aspöck, L., Pausch, F., Stienen, J., Berzborn, M., Michael, K., Fels, J., & Vorländer, M. (2018a). Application of virtual acoustic environments in the scope of auditory research [Invited paper.]. *XXVIII Encontro da Sociedade Brasileira de Acústica (SOBRAC 2018).* https://doi.org/10.17648/sobrac-87162

MacCutcheon, D., Pausch, F., Fels, J., & Ljung, R. (2017). The relationship between working memory and second language speech reception thresholds in sequential bilingual children. *APCAM 2017: 16th Annual Auditory Perception, Cognition, and Action Meeting*, 14–14.

Pausch, F., & Aspöck, L. (2015). Real-time acoustic scene rendering to investigate auditory and cognitive aspects of hearing impaired subjects [Bronze award in the best poster competition.]. *19th International Student Conference on Electrical Engineering*.

Pausch, F., Aspöck, L., Vorländer, M., & Fels, J. (2018c). A hybrid real-time auralization system for binaural reproduction of virtual acoustic environments with simulated room acoustics adapted for hearing aid users. *The Journal of the Acoustical Society of America*, *143*(3), 1823–1823. https://doi.org/10.1121/1.5035989

Pausch, F., Peng, Z. E., Aspöck, L., & Fels, J. (2016b). Influence of room acoustics on speech perception by children in a dynamic virtual acoustic environment with simulated hearing aids. *Hearing Across the Lifespan*.

Pausch, F., Peng, Z. E., Aspöck, L., & Fels, J. (2016c). Speech perception by children in a real-time virtual acoustic environment with simulated hearing aids and room acoustics [Invited paper.]. *22nd International Congress on Acoustics: ICA 2016*.

Pausch, F., Pomberger, H., & Zotter, F. (2013). A New Double-Cone Microphone Array for High-Definition Panoramic Audio Recording [Gold award in the student design competition (graduate category).]. *134th Conference of the Audio Engineering Society (AES)*.

Peng, Z. E., Pausch, F., Aspöck, L., & Fels, J. (2016). Effect of Reverberation on Spatial Processing by Children with Hearing Loss or ADHD. *Audiology Now*, PP1018.

Peng, Z. E., Pausch, F., & Fels, J. (2016b). Effect of room acoustics on speech perception by children with hearing loss. *The Journal of the Acoustical Society of America*, *139*(4), 1979–1979. https://doi.org/10.1121/1.4949774

Peng, Z. E., Pausch, F., & Fels, J. (2018). Spatial release from masking in reverberation for children and adults with normal hearing. *The Journal of the Acoustical Society of America*, *144*(3), 1892–1893. https://doi.org/10.1121/1.5068297

Contributions to national conferences

Aspöck, L., Pausch, F., & Fels, J. (2015). Echtzeit-Simulation und Wiedergabe komplexer Schallfelder für Hörgeräteträger. *46th Annual Meeting of the German Society for Medical Physics, Marburg, 9-12 September*.

Aspöck, L., Pausch, F., Vorländer, M., & Fels, J. (2015). Dynamic real-time auralization for experiments on the perception of hearing impaired subjects. *German Annual Conference on Acoustics: DAGA 2015, Nuremberg, 16-19 March*, 1621–1624.

Braren, H. S., Dreier, C. G., Pausch, F., Vorländer, M., & Fels, J. (2020). Prediction of Diesel Roughness with TPA. *R596*.

Dreier, C. G., Braren, H. S., Pausch, F., Fels, J., & Vorländer, M. (2019). Prediction of Diesel Roughness Using Binaural Transfer Path Synthesis. *11th Aachen Acoustics Colloquium (AAC 2019): Aachen, November 25-27, 2019*.

Fels, J., Bomhardt, R., & Pausch, F. (2016). Investigation on Localization Performance Using Smoothed Individual Head-Related Transfer Functions. *42th German Annual Conference on Acoustics: DAGA 2016, Aachen, 14-17 March*, 86–88.

Fels, J., Bomhardt, R., & Pausch, F. (2017). Subjektive Wahrnehmungsschwellen geglät-
teter Phasenspektren von Außenohrübertragungsfunktionen. *43th German Annual
Conference on Acoustics: DAGA 2017, Kiel, 6-9 March.*

Loh, K., Pausch, F., & Fels, J. (2018). Classification of Rooms in Educational Buildings
Using Different Noise Indicators. *44th German Annual Conference on Acoustics:
DAGA 2018, Munich, 19-22 March.*

Pausch, F., Aspöck, L., & Fels, J. (2015). Evaluation of binaural sound reproduction
systems with focus on perceptual quality. *42th German Annual Conference on
Acoustics: DAGA 2015, Nuremberg, 16-19 March*, 1456–1459.

Pausch, F., & Fels, J. (2017). Auditory distance perception in virtual acoustic environments using binaural technology and simulated room acoustics. *43th German
Annual Conference on Acoustics: DAGA 2017, Kiel, 6-9 March.*

Pausch, F., & Fels, J. (2018b). Speech reception thresholds of children and adults with
hearing loss, ADHD and CAPD in a virtual acoustic environment. *44th German
Annual Conference on Acoustics: DAGA 2018, Munich, 19-22 March.*

Pausch, F., Kohnen, M., Aspöck, L., Peng, Z. E., & Fels, J. (2016a). Investigation
of sound localization performance in a virtual acoustic environment designed for
hearing aid users. *42th German Annual Conference on Acoustics: DAGA 2016,
Aachen, 14-17 March*, 1124–1127.

Pausch, F., Kohnen, M., Stienen, J., Aspöck, L., Vorländer, M., & Fels, J. Implementation and application of acoustic crosstalk cancellation systems. In: In *Precolloquium
"Spatial Audio Reproduction Systems" (Organiser: Sascha Spors), 45th German
Annual Conference on Acoustics: DAGA 2019, Rostock, 18-21 March.* Rostock,
2019, March 18. https://publications.rwth-aachen.de/record/774156

Pausch, F., Vollmer, L., Kohnen, M., & Fels, J. (2020). Comparison of Error Metrics
in Loudspeaker-Based Reproduction Systems. *46th German Annual Conference on
Acoustics: DAGA 2020, Hannover, 16-19 March.*

Peng, Z. E., Pausch, F., & Fels, J. (2016a). Auditory Training of Spatial Processing
in Children with Hearing Loss in Virtual Acoustic Environments: Pretest Results.
42th German Annual Conference on Acoustics: DAGA 2016, Aachen, 14-17 March,
1128–1131.

Tumbrägel, J., Pausch, F., Aspöck, L., & Fels, J. (2016). Investigations on binaural
reproduction in virtual acoustic environments using hearing aids. *42th German
Annual Conference on Acoustics: DAGA 2016, Aachen, 14-17 March.*

Viveros, R., Peng, Z. E., Pausch, F., & Fels, J. (2016). Effect of a moving distractor on
speech intelligibility in babble noise using a digit-triplet test. *42th German Annual
Conference on Acoustics: DAGA 2016, Aachen, 14-17 March*, 673–676.

Invited talks

Aspöck, L., Pausch, F., Vorländer, M., & Fels, J. (2018b). Integration of computer
models to simulate and evaluate hearing aids. *Colloquium at Laboratório de Vibrações
e Acústica, Universidade Federal de Santa Catarina, Florianópolis, Brazil.*

Pausch, F. (2018). How we perceive music in concert halls - An introduction to acoustics for musicians. *Colloqium at Department of Music, University of Pretoria, South Africa.*

Pausch, F., Aspöck, L., Vorländer, M., & Fels, J. (2018a). Making virtual acoustic environments accessible for hearing aid users. *Colloqium at Laboratório de Vibrações e Acústica, Universidade Federal de Santa Catarina, Florianópolis, Brazil.*

Pausch, F., & Fels, J. (2018a). A binaural real-time reproduction system to investigate speech perception of subjects with hearing loss under simulated room acoustics. *Colloqium at Department of Speech-Language Pathology and Audiology, University of Pretoria, South Africa.*

Pausch, F., Fels, J., & Aspöck, L. (2016). Trends in Acoustic Virtual Reality: How can we extend virtual acoustic environments for hearing aid users? *Colloqium at Technical University of Denmark, Lyngby/Copenhagen (Denmark).*

Supervised student work

Loh, K. (2017, May). *Investigation of noise exposure in educational institutions using aurally accurate techniques for preschool and primary school-aged children* (Master's thesis). Teaching and Research Area of Medical Acoustics, Institute of Technical Acoustics, RWTH Aachen University, Germany.

Neuhöfer, W. (2018, December). *Roughness and overall sound quality perception of diesel engines depending on different playback methods* (Bachelor's thesis). Teaching and Research Area of Medical Acoustics, Institute of Technical Acoustics, RWTH Aachen University, Germany.

Stockmann, A. (2015, December). *Investigation of realistic acoustic scene simulation for people with hearing impairment* (Master's thesis). Teaching and Research Area of Medical Acoustics, Institute of Technical Acoustics, RWTH Aachen University, Germany.

Tumbrägel, J. (2015, September). *Hearing aid signal processing for a real-time auralization system* (Master's thesis). Teaching and Research Area of Medical Acoustics, Institute of Technical Acoustics, RWTH Aachen University, Germany.

Vollmer, L. (2016, November). *Perceptual influence of head-related transfer functions with smoothed phase spectra on the localization of virtual sound sources* (Bachelor's thesis) [Study prize by the Association for Technical Acoustics eV (GfTA) for the best bachelor's thesis.]. Teaching and Research Area of Medical Acoustics, Institute of Technical Acoustics, RWTH Aachen University, Germany.

Vollmer, L. (2018, December). *Evaluation of loudspeaker-based spatial audio reproduction methods for hearing aid applications* (Master's thesis) [Study prize by the Association for Technical Acoustics eV (GfTA) for the best master's thesis.]. Teaching and Research Area of Medical Acoustics, Institute of Technical Acoustics, RWTH Aachen University, Germany.

Bibliography

Aaronson, N. L., & Hartmann, W. M. (2014). Testing, correcting, and extending the woodworth model for interaural time difference. *The Journal of the Acoustical Society of America*, *135*(2), 817–823. https://doi.org/10.1121/1.4861243

Abramovitz, M., & Stegun, I. A. (1964). *Handbook of Mathematical Functions: With Formulas, Graphs, and Mathematical Tables*. Dover.

Adiloğlu, K., Kayser, H., Baumgärtel, R. M., Rennebeck, S., Dietz, M., & Hohmann, V. (2015). A Binaural Steering Beamformer System for Enhancing a Moving Speech Source [PMID: 26721924]. *Trends in Hearing*, *19*, 2331216515618903. https://doi.org/10.1177/2331216515618903

Agarwal, R., & Burrus, C. (1974). Fast convolution using Fermat number transforms with applications to digital filtering. *IEEE Transactions on Acoustics, Speech, and Signal Processing*, *22*(2), 87–97.

Ahrens, A., Lund, K. D., Marschall, M., & Dau, T. (2019). Sound source localization with varying amount of visual information in virtual reality. *PLOS ONE*, *14*(3), 1–19. https://doi.org/10.1371/journal.pone.0214603

Akeroyd, M. A., Chambers, J., Bullock, D., Palmer, A. R., Summerfield, A. Q., Nelson, P. A., & Gatehouse, S. (2007). The binaural performance of a cross-talk cancellation system with matched or mismatched setup and playback acoustics. *The Journal of the Acoustical Society of America*, *121*(2), 1056–1069. https://doi.org/10.1121/1.2404625

Albrecht, R., & Lokki, T. (2013). Adjusting the perceived distance of virtual speech sources by modifying binaural room impulse responses. *International Conference on Auditory Display (ICAD 2013), Lodz, Poland, July 6-10, 2013*, 233–241.

Algazi, V. R., Avendano, C., & Duda, R. O. (2001). Estimation of a Spherical-Head Model from Anthropometry. *Journal of the Audio Engineering Society*, *49*(6), 472–479.

Algazi, V. R., Duda, R. O., Thompson, D. M., & Avendano, C. (2001). The CIPIC HRTF database. *Applications of Signal Processing to Audio and Acoustics, 2001 IEEE Workshop on the*, 99–102. https://doi.org/10.1109/ASPAA.2001.969552

Allen, J. B., & Berkley, D. A. (1979). Image method for efficiently simulating small-room acoustics. *The Journal of the Acoustical Society of America*, *65*(4), 943–950. https://doi.org/10.1121/1.382599

Anderson, P. W., & Zahorik, P. (2014). Auditory/visual distance estimation: Accuracy and variability. *Frontiers in psychology*, *5*. https://doi.org/10.3389/fpsyg.2014.01097

Andreopoulou, A., & Katz, B. F. G. (2017). Identification of perceptually relevant methods of inter-aural time difference estimation. *The Journal of the Acoustical Society of America*, *142*(2), 588–598. https://doi.org/10.1121/1.4996457

ANSI/ASA S12.60-2010/Part 1. (2010). *Acoustical Performance Criteria, Design Requirements, and Guidelines for Schools, Part 1: Permanent Schools* (Norm ANSI/ASA S12.60-2010/Part 1). American National Standards Institute. Washington, D.C., United States.

ANSI/ASA S3.1. (1999). *Maximum Permissible Ambient Noise Levels for Audiometric Test Rooms* (Norm ANSI/ASA S3.1). American National Standards Institute. Washington, D.C., United States.

Arlinger, S. D. (1979). Comparison of Ascending and Bracketing Methods in Pure Tone Audiometry A Multi-laboratory Study [PMID: 531479]. *Scandinavian Audiology*, *8*(4), 247–251. https://doi.org/10.3109/01050397909076327

Aspöck, L., Pelzer, S., Wefers, F., & Vorländer, M. (2014). A Real-Time Auralization Plugin for Architectural Design and Education. *Proceedings of the EAA Joint Symposium on Auralization and Ambisonics 2014*, 156–161. https://doi.org/10.14279/depositonce-26

Atal, B. S., Hill, M., & Schroeder, M. R. (1966). Apparent sound source translator [US Patent 3,236,949].

Axelsson, A., & Sandh, A. (1985). Tinnitus in noise-induced hearing loss [PMID: 4074979]. *British Journal of Audiology*, *19*(4), 271–276. https://doi.org/10.3109/03005368509078983

Bai, M. R., & Lee, C.-C. (2006). Objective and subjective analysis of effects of listening angle on crosstalk cancellation in spatial sound reproduction. *The Journal of the Acoustical Society of America*, *120*(4), 1976–1989. https://doi.org/10.1121/1.2257986

Baiduc, R. R., Poling, G. L., Hong, O., & Dhar, S. (2013). Clinical measures of auditory function: The cochlea and beyond. *Disease-a-month: DM*, *59*(4), 147. https://doi.org/10.1016/j.disamonth.2013.01.005

Bates, D., Mächler, M., Bolker, B., & Walker, S. (2015). Fitting Linear Mixed-Effects Models Using lme4. *Journal of Statistical Software*, *67*(1), 1–48. https://doi.org/10.18637/jss.v067.i01

Batke, J.-M., & Keiler, F. (2010). Using VBAP-derived panning functions for 3D ambisonics decoding. *2nd Ambisonics Symposium, Paris*.

Bauck, J., & Cooper, D. H. (1996). Generalized Transaural Stereo and Applications. *Journal of the Audio Engineering Society*, *44*(9), 683–705.

Bauer, B. B. (1961). Stereophonic earphones and binaural loudspeakers. *Journal of the Audio Engineering Society*, *9*(2), 148–151.

Baumgartner, R., & Majdak, P. (2015). Modeling Localization of Amplitude-Panned Virtual Sources in Sagittal Planes. *Journal of the Audio Engineering Society. Audio Engineering Society, 63*(7-8), 562.

Baumgartner, R., Majdak, P., & Laback, B. (2014). Modeling sound-source localization in sagittal planes for human listeners. *The Journal of the Acoustical Society of America, 136*(2), 791–802. https://doi.org/10.1121/1.4887447

Begault, D. R., McClain, B. U., & Anderson, M. R. (2004). Early reflection thresholds for anechoic and reverberant stimuli within a 3-D sound display. *Proc. 18th Int. Congress on Acoust.(ICA04), Kyoto, Japan.*

Begault, D. R., Wenzel, E. M., & Anderson, M. R. (2001). Direct comparison of the impact of head tracking, reverberation, and individualized head-related transfer functions on the spatial perception of a virtual speech source. *Journal of the Audio Engineering Society, 49*(10), 904–916.

Bentler, R., & Chiou, L.-K. (2006). Digital Noise Reduction: An Overview [PMID: 16959731]. *Trends in Amplification, 10*(2), 67–82. https://doi.org/10.1177/1084713806289514

Bernstein, L. R., & Trahiotis, C. (1985). Lateralization of low-frequency, complex waveforms: The use of envelope-based temporal disparities. *The Journal of the Acoustical Society of America, 77*(5), 1868–1880. https://doi.org/10.1121/1.391938

Bertet, S., Daniel, J., & Moreau, S. 3D Sound Field Recording with Higher Order Ambisonics - Objective Measurements and Validation of Spherical Microphone. In: In *Audio engineering society convention 120.* 2006, May. http://www.aes.org/e-lib/browse.cfm?elib=13661

Bertoli, S., Staehelin, K., Zemp, E., Schindler, C., Bodmer, D., & Probst, R. (2009). Survey on hearing aid use and satisfaction in Switzerland and their determinants [PMID: 19363719]. *International Journal of Audiology, 48*(4), 183–195. https://doi.org/10.1080/14992020802572627

Berzborn, M., Bomhardt, R., Klein, J., Richter, J.-G., & Vorländer, M. (2017). The ITA-Toolbox: An Open Source MATLAB Toolbox for Acoustic Measurements and Signal Processing. *43th Annual German Congress on Acoustics, Kiel (Germany), 6 Mar 2017 - 9 Mar 2017,* 222–225.

Best, V., Kalluri, S., McLachlan, S., Valentine, S., Edwards, B., & Carlile, S. (2010). A comparison of CIC and BTE hearing aids for three-dimensional localization of speech. *International Journal of Audiology, 49*(10), 723–732. https://doi.org/10.3109/14992027.2010.484827

Best, V., Mason, C. R., & Kidd, G. (2011). Spatial release from masking in normally hearing and hearing-impaired listeners as a function of the temporal overlap of competing talkers. *The Journal of the Acoustical Society of America, 129*(3), 1616–1625. https://doi.org/10.1121/1.3533733

Best, V., Mejia, J., Freeston, K., van Hoesel, R. J., & Dillon, H. (2015). An evaluation of the performance of two binaural beamformers in complex and dynamic multitalker environments [PMID: 26140298]. *International Journal of Audiology, 54*(10), 727–735. https://doi.org/10.3109/14992027.2015.1059502

Bisgaard, N., & Ruf, S. (2017). Findings From EuroTrak Surveys From 2009 to 2015: Hearing Loss Prevalence, Hearing Aid Adoption, and Benefits of Hearing Aid Use. *American Journal of Audiology, 26*(3S), 451–461. https://doi.org/10.1044/2017\ _AJA-16-0135

Blauert, J. (1997). *Spatial Hearing: The Psychophysics of Human Sound Localization.* Cambridge, Mass. MIT Press.

Blauert, J., Mourjopoulos, J., & Buchholz, J. Room Masking: Understanding and Modelling the Masking of Reflections in Rooms. In: In *Audio engineering society convention 110.* 2001, May. http://www.aes.org/e-lib/browse.cfm?elib=9931

Boersma, P., & Weenink, D. (2016, May). PRAAT: Doing phonetics by computer. Version 6.0.18 [Available at https://praat.de.uptodown.com/windows/versions, accessed on 2021-08-23].

Boi, R., Racca, L., Cavallero, A., Carpaneto, V., Racca, M., Dall' Acqua, F., Ricchetti, M., Santelli, A., & Odetti, P. (2012). Hearing loss and depressive symptoms in elderly patients. *Geriatrics & Gerontology International, 12*(3), 440–445. https://doi.org/https://doi.org/10.1111/j.1447-0594.2011.00789.x

Bomhardt, R. (2017). *Anthropometric individualization of head-related transfer functions analysis and modeling* (Doctoral dissertation). Teaching and Research Area of Medical Acoustics, RWTH Aachen University. Aachen, Germany, Logos Verlag Berlin GmbH.

Bomhardt, R., de la Fuente Klein, M., & Fels, J. (2016). A high-resolution head-related transfer function and three-dimensional ear model database. *Proceedings of Meetings on Acoustics, 29*(1), 050002. https://doi.org/10.1121/2.0000467

Braat-Eggen, P. E., van Heijst, A., Hornikx, M., & Kohlrausch, A. (2017). Noise disturbance in open-plan study environments: A field study on noise sources, student tasks and room acoustic parameters [PMID: 28287041]. *Ergonomics, 60*(9), 1297–1314. https://doi.org/10.1080/00140139.2017.1306631

Bregman, A. S. (1994). *Auditory scene analysis: The perceptual organization of sound.* MIT Press.

Brinkmann, F., Lindau, A., Weinzierl, S., Par, S. v. d., Müller-Trapet, M., Opdam, R., & Vorländer, M. (2017). A High Resolution and Full-Spherical Head-Related Transfer Function Database for Different Head-Above-Torso Orientations. *Journal of the Audio Engineering Society, 65*(10), 841–848.

Brinkmann, F., Roden, R., Lindau, A., & Weinzierl, S. (2015). Audibility and Interpolation of Head-Above-Torso Orientation in Binaural Technology. *IEEE Journal of Selected Topics in Signal Processing, 9*(5), 931–942. https://doi.org/10.1109/JSTSP.2015.2414905

Bronkhorst, A. W., & Plomp, R. (1992). Effect of multiple speechlike maskers on binaural speech recognition in normal and impaired hearing. *The Journal of the Acoustical Society of America, 92*(6), 3132–3139. https://doi.org/10.1121/1.404209

Bronkhorst, A. W. (1995). Localization of real and virtual sound sources. *The Journal of the Acoustical Society of America, 98*(5), 2542–2553. https://doi.org/10.1121/1.413219

Bronkhorst, A. W. (2000). The Cocktail Party Phenomenon: A Review of Research on Speech Intelligibility in Multiple-Talker Conditions. *Acta Acustica united with Acustica, 86*(1), 117–128.

Bronkhorst, A. W., & Plomp, R. (1990). A clinical test for the assessment of binaural speech perception in noise. *Audiology, 29*(5), 275–285. https://doi.org/10.3109/00206099009072858

Brungart, D. S., Simpson, B. D., & Kordik, A. J. (2005). The detectability of headtracker latency in virtual audio displays. *Proceedings of the 11th International Conference on Auditory Display (ICAD2005)*, 37–42.

Brungart, D. S. (2001). Informational and energetic masking effects in the perception of two simultaneous talkers. *The Journal of the Acoustical Society of America, 109*(3), 1101–1109. https://doi.org/10.1121/1.1345696

Brungart, D. S., & Rabinowitz, W. M. (1999). Auditory localization of nearby sources. Head-related transfer functions. *The Journal of the Acoustical Society of America, 106*(3), 1465–1479. https://doi.org/10.1121/1.427180

Byrne, D., Dillon, H., Ching, T., Katsch, R., & Keidser, G. (2001). NAL-NL1 procedure for fitting nonlinear hearing aids: characteristics and comparisons with other procedures. *Journal of the American academy of audiology, 12*(1).

Byrne, D., Noble, W., & Glauerdt, B. (1996). Effects of earmold type on ability to locate sounds when wearing hearing aids. *Ear and hearing, 17*(3), 218–228. https://doi.org/10.1097/00003446-199606000-00005

Cameron, S., & Dillon, H. (2007). Development of the Listening in Spatialized Noise-Sentences Test (LISN-S). *Ear and hearing, 28*(2), 196–211.

Cameron, S., & Dillon, H. (2011). Development and Evaluation of the LiSN & Learn Auditory Training Software for Deficit-Specific Remediation of Binaural Processing Deficits in Children: Preliminary Findings. *Journal of the American Academy of Audiology, 22*(10), 678–696. https://doi.org/10.3766/jaaa.22.10.6

Cameron, S., Dillon, H., & Newall, P. (2006). The Listening in Spatialized Noise Test: An Auditory Processing Disorder Study. *Journal of the American Academy of Audiology, 17*(5), 306–320. https://doi.org/10.3766/jaaa.17.5.2

Cameron, S., Glyde, H., & Dillon, H. (2011). Listening in Spatialized Noise-Sentences Test (LiSN-S): normative and retest reliability data for adolescents and adults up to 60 years of age. *Journal of the American Academy of Audiology, 22*(10), 697–709.

Cameron, S., Glyde, H., & Dillon, H. (2012). Efficacy of the LiSN & Learn auditory training software: randomized blinded controlled study. *Audiology research, 2*(1), 86–93. https://doi.org/10.4081/audiores.2012.e15

Chen, F. (2003). Localization of 3-D sound presented through headphone-Duration of sound presentation and localization accuracy. *Journal of the Audio Engineering Society, 51*(12), 1163–1171.

Cherry, E. C. (1953). Some Experiments on the Recognition of Speech, with One and with Two Ears. *The Journal of the Acoustical Society of America, 25*(5), 975–979. https://doi.org/10.1121/1.1907229

Ching, T. Y. C., van Wanrooy, E., Dillon, H., & Carter, L. (2011). Spatial release from masking in normal-hearing children and children who use hearing aids. *The Journal of the Acoustical Society of America*, *129*(1), 368–375. https://doi.org/10.1121/1.3523295

Chisolm, T. H., Johnson, C. E., Danhauer, J. L., Portz, L. J., Abrams, H. B., Lesner, S., McCarthy, P. A., & Newman, C. W. (2007). A Systematic Review of Health-Related Quality of Life and Hearing Aids: Final Report of the American Academy of Audiology Task Force on the Health-Related Quality of Life Benefits of Amplification in Adults. *Journal of the American Academy of Audiology*, *18*(02), 151–183.

Choueiri, E. Y. (2008). Optimal Crosstalk Cancellation for Binaural Audio with Two Loudspeakers [Available at http://www.princeton.edu/3D3A/Publications/BACCHPaperV4d.pdf]. *Princeton University*, *28*.

Clark, J. G., et al. (1981). Uses and abuses of hearing loss classification. *Asha*, *23*(7), 493–500.

Clifton, R. K., Gwiazda, J., Bauer, J. A., Clarkson, M. G., & Held, R. M. (1988). Growth in head size during infancy: Implications for sound localization. *Developmental Psychology*, *24*(4), 477. https://doi.org/10.1037/0012-1649.24.4.477

Compton-Conley, C. L., Neuman, A. C., Killion, M. C., & Levitt, H. (2004). Performance of Directional Microphones for Hearing Aids: Real-World versus Simulation. *Journal of the American Academy of Audiology*, *15*(6), 440–455. https://doi.org/10.3766/jaaa.15.6.5

Cord, M., Baskent, D., Kalluri, S., & Moore, B. C. (2007). Disparity between clinical assessment and real-world performance of hearing aids. *Journal of the American Academy of Audiology*, *14*(6), 22. https://doi.org/10.1055/s-0040-1715973

Cord, M. T., Surr, R. K., Walden, B. E., & Olson, L. (2002). Performance of Directional Microphone Hearing Aids in Everyday Life. *Journal of the American Academy of Audiology*, *13*(6), 295–307.

Cornelisse, L. E., Seewald, R. C., & Jamieson, D. G. (1995). The input/output formula: A theoretical approach to the fitting of personal amplification devices. *The Journal of the Acoustical Society of America*, *97*(3), 1854–1864. https://doi.org/10.1121/1.412980

Crandell, C. C., & Smaldino, J. (2007). *Room acoustics for listeners with normal-hearing and hearing impairment* (R. R. Michael Valente Holly Hosford-Dunn, Ed.; 2nd edition). Thieme New York.

Crandell, C. C., & Smaldino, J. J. (2000). Classroom Acoustics for Children With Normal Hearing and With Hearing Impairment. *Language, Speech, and Hearing Services in Schools*, *31*(4), 362–370. https://doi.org/10.1044/0161-1461.3104.362

Cubick, J., & Dau, T. (2016). Validation of a Virtual Sound Environment System for Testing Hearing Aids. *Acta Acustica united with Acustica*, *102*(3), 547–557. https://doi.org/10.3813/AAA.918972

Culling, J. F. (2016). Speech intelligibility in virtual restaurants. *The Journal of the Acoustical Society of America*, *140*(4), 2418–2426. https://doi.org/10.1121/1.4964401

Culling, J. F., & Colburn, H. S. (2000). Binaural sluggishness in the perception of tone sequences and speech in noise. *The Journal of the Acoustical Society of America, 107*(1), 517–527. https://doi.org/10.1121/1.428320

Curran, J. R., & Galster, J. A. (2013). The Master Hearing Aid. *Trends in Amplification, 17*(2), 108–134. https://doi.org/10.1177/1084713813486851

Daniel, J. (2000). *Représentation de champs acoustiques, application à la transmission et à la reproduction de scènes sonores complexes dans un contexte multimédia* (Doctoral dissertation). University of Paris VI, France.

David, D., & Werner, P. (2016). Stigma regarding hearing loss and hearing aids: A scoping review. *Stigma and Health, 1*(2), 59. https://doi.org/https://doi.org/10.1037/sah0000022

Davis, T. J., Grantham, D. W., & Gifford, R. H. (2016). Effect of motion on speech recognition. *Hearing research, 337*, 80–88. https://doi.org/10.1016/j.heares.2016.05.011

Delaunay, B., et al. (1934). Sur la sphere vide. *Izv. Akad. Nauk SSSR, Otdelenie Matematicheskii i Estestvennyka Nauk, 7*(793-800), 1–2.

Denk, F., Ernst, S. M. A., Ewert, S. D., & Kollmeier, B. (2018). Adapting Hearing Devices to the Individual Ear Acoustics: Database and Target Response Correction Functions for Various Device Styles [PMID: 29877161]. *Trends in Hearing, 22*, 2331216518779313. https://doi.org/10.1177/2331216518779313

Denk, F., Ewert, S. D., & Kollmeier, B. (2018). Spectral directional cues captured by hearing device microphones in individual human ears. *The Journal of the Acoustical Society of America, 144*(4), 2072–2087. https://doi.org/10.1121/1.5056173

Denk, F., & Kollmeier, B. (2021). The Hearpiece database of individual transfer functions of an in-the-ear earpiece for hearing device research. *Acta Acust., 5*, 2. https://doi.org/10.1051/aacus/2020028

Diebel, J. (2006). Representing attitude: Euler angles, unit quaternions, and rotation vectors. *Matrix, 58*(15-16), 1–35.

Dietrich, P., Masiero, B., & Vorländer, M. (2013). On the Optimization of the Multiple Exponential Sweep Method. *Journal of the Audio Engineering Society, 61*(3), 113–124.

Dillon, H. (2012). *Hearing Aids* (2nd edition). Boomerang Press; Thieme Sydney: New York.

DIN 18041. (2016, March). *Acoustic quality in rooms - Specifications and instructions for the room acoustic design* (Standard DIN 18041:2016-03). German Institute for Standardization, e.V. Berlin, Germany.

DIN 45631. (1991, March). *Procedure for calculating loudness level and loudness* (Norm DIN45631). German Institute for Standardisation. Berlin, Germany.

DIN 45641:1990-06. (1990, June). *Mittelung von Schallpegeln (Averaging of sound levels)* (Norm DIN 45641:1990-06). German Institute for Standardization, e.V. Berlin, Germany.

Dreier, C., & Vorländer, M. (2021). Aircraft noise–Auralization-based assessment of weather-dependent effects on loudness and sharpness. *The Journal of the Acoustical Society of America*, *149*(5), 3565–3575. https://doi.org/10.1121/10.0005040

Ehret, J., Stienen, J., Brozdowski, C., Bönsch, A., Mittelberg, I., Vorländer, M., & Kuhlen, T. W. (2020). Evaluating the Influence of Phoneme-Dependent Dynamic Speaker Directivity of Embodied Conversational Agents' Speech. *Proceedings of the 20th ACM International Conference on Intelligent Virtual Agents*, 1–8. https://doi.org/10.1145/3383652.3423863

European Union. (2016). Regulation (EU) 2016/679 of the European Parliament and of the Council of 27 April 2016 on the protection of natural persons with regard to the processing of personal data and on the free movement of such data, and repealing Directive 95/46/EC (General Data Protection Regulation). *Official Journal L110*, *59*, 1–88.

Evans, W., Dyreby, J., Bech, S., Zielinski, S., & Rumsey, F. Effects of Loudspeaker Directivity on Perceived Sound Quality - A Review of Existing Studies. In: In *Audio engineering society convention 126*. 2009, May. http://www.aes.org/e-lib/browse.cfm?elib=14941

Farina, A. Simultaneous Measurement of Impulse Response and Distortion with a Swept-Sine Technique. In: In *Audio engineering society convention 108*. 2000, February. http://www.aes.org/e-lib/browse.cfm?elib=10211

Farina, A. Advancements in Impulse Response Measurements by Sine Sweeps. In: In *Audio engineering society convention 122*. 2007, May. http://www.aes.org/e-lib/browse.cfm?elib=14106

Fellgett, P. (1975). Ambisonics. Part One: General System Description. *Studio Sound*, *17*(8), 20–22.

Fels, J., Buthmann, P., & Vorländer, M. (2004). Head-related transfer functions of children. *Acta Acustica united with Acustica*, *90*(5), 918–927.

Fels, J. (2006). Ear canal impedances of children and adults–investigations with simulation and measurements. *Acta Otorrinolaringol Esp*, *57*, 82–89.

Fels, J. (2008). *From Children to Adults: How Binaural Cues and Ear Canal Impedances Grow.*

Finitzo-Hieber, T., & Tillman, T. W. (1978). Room Acoustics Effects on Monosyllabic Word Discrimination Ability for Normal and Hearing-Impaired Children. *Journal of Speech and Hearing Research*, *21*(3), 440–458. https://doi.org/10.1044/jshr.2103.440

Frank, M. (2013). *Phantom sources using multiple loudspeakers in the horizontal plane* (Doctoral dissertation). Institute of Electronic Music, Acoustics, University of Music, and Performing Arts Graz, Austria.

Frank, M., & Zotter, F. Exploring the Perceptual Sweet Area in Ambisonics. In: In *Audio engineering society convention 142*. 2017, May. http://www.aes.org/e-lib/browse.cfm?elib=18604

Frank, M., Zotter, F., & Sontacchi, A. (2008). Localization Experiments Using Different 2D Ambisonics Decoders. *25th Tonmeistertagung, VDT International Convention*, 1–9.

193

Friston, S., & Steed, A. (2014). Measuring Latency in Virtual Environments. *IEEE Transactions on Visualization and Computer Graphics*, *20*(4), 616–625. https://doi.org/10.1109/TVCG.2014.30

Fuglsang, S. A., Dau, T., & Hjortkjær, J. (2017). Noise-robust cortical tracking of attended speech in real-world acoustic scenes. *NeuroImage*, *156*, 435–444. https://doi.org/10.1016/j.neuroimage.2017.04.026

Gardner, M. B. (1968). Historical Background of the Haas and/or Precedence Effect. *The Journal of the Acoustical Society of America*, *43*(6), 1243–1248. https://doi.org/10.1121/1.1910974

Gardner, W. G., & Martin, K. D. (1995). HRTF measurements of a KEMAR. *The Journal of the Acoustical Society of America*, *97*(6), 3907–3908. https://doi.org/10.1121/1.412407

General Assembly of the World Medical Association and others. (2014). World Medical Association Declaration of Helsinki: ethical principles for medical research involving human subjects. *The Journal of the American College of Dentists*, *81*(3), 14–18.

Georganti, E., May, T., van de Par, S., & Mourjopoulos, J. (2013). Extracting Sound-Source-Distance Information from Binaural Signals. In J. Blauert (Ed.), *The technology of binaural listening* (pp. 171–199). Springer Berlin Heidelberg. https://doi.org/10.1007/978-3-642-37762-4_7

Gerzon, M. A. (1975). Ambisonics. Part Two: Studio Techniques. *Studio Sound*, *17*(8), 24–26.

Gerzon, M. A. (1985). Ambisonics in Multichannel Broadcasting and Video. *Journal of the Audio Engineering Society*, *33*(11), 859–871.

Gerzon, M. A. General Metatheory of Auditory Localisation. In: In *Audio engineering society convention 92*. 1992, March. http://www.aes.org/e-lib/browse.cfm?elib=6827

Gilkey, R., & Anderson, T. R. (2014). *Binaural and Spatial Hearing in Real and Virtual Environments*. Taylor & Francis.

Glyde, H., Cameron, S., Dillon, H., Hickson, L., & Seeto, M. (2013). The effects of hearing impairment and aging on spatial processing. *Ear and hearing*, *34*(1), 15–28. https://doi.org/10.1097/AUD.0b013e3182617f94

Gnewikow, D., Ricketts, T., Bratt, G. W., & Mutchler, L. C. (2009). Real-world benefit from directional microphone hearing aids. *Journal of Rehabilitation Research & Development*, *46*(5), 603–618. https://doi.org/10.1682/jrrd.2007.03.0052

Gogel, W. C. (1969). The sensing of retinal size. *Vision Research*, *9*(9), 1079–1094. https://doi.org/https://doi.org/10.1016/0042-6989(69)90049-2

Goldstein, H., Poole, C., & Safko, J. (2002). *Classical Mechanics*. American Association of Physics Teachers.

Green, D. M. (1993). A maximum-likelihood method for estimating thresholds in a yes–no task. *The Journal of the Acoustical Society of America*, *93*(4), 2096–2105. https://doi.org/10.1121/1.406696

Grimm, G., Ewert, S., & Hohmann, V. (2015). Evaluation of spatial audio reproduction schemes for application in hearing aid research. *Acta Acustica united with Acustica*, *101*(4), 842–854. https://doi.org/10.3813/AAA.918878

Grimm, G., Herzke, T., Berg, D., & Hohmann, V. (2006). The master hearing aid: A PC-based platform for algorithm development and evaluation. *Acta acustica united with Acustica*, *92*(4), 618–628.

Grimm, G., Kollmeier, B., & Hohmann, V. (2016). Spatial Acoustic Scenarios in Multi-channel Loudspeaker Systems for Hearing Aid Evaluation. *Journal of the American Academy of Audiology*, *27*(7), 557–566. https://doi.org/10.3766/jaaa.15095

G. T. M. (1932). Long Lines On The Earth. *Empire Survey Review*, *1*(6), 259–263. https://doi.org/10.1179/sre.1932.1.6.259

Guang, P., Fu, Z., Xie, L., & Zhao, W. (2016). Study on near-field crosstalk cancellation based on least square algorithm. *Signal and Information Processing Association Annual Summit and Conference (APSIPA), 2016 Asia-Pacific*, 1–5. https://doi.org/10.1109/APSIPA.2016.7820737

Gurgel, R. K., Jackler, R. K., Dobie, R. A., & Popelka, G. R. (2012). A New Standardized Format for Reporting Hearing Outcome in Clinical Trials [PMID: 22931898]. *Otolaryngology–Head and Neck Surgery*, *147*(5), 803–807. https://doi.org/10.1177/0194599812458401

Guski, M. (2015). *Influences of external error sources on measurements of room acoustic parameters* (Doctoral dissertation). Institute of Technical Acoustics, RWTH Aachen University, Germany. Aachen, Logos-Verlag.

Halekoh, U., & Højsgaard, S. (2014). A Kenward-Roger Approximation and Parametric Bootstrap Methods for Tests in Linear Mixed Models – The R Package pbkrtest. *Journal of Statistical Software*, *59*(9), 1–30.

Halkosaari, T., Vaalgamaa, M., & Karjalainen, M. (2005). Directivity of artificial and human speech. *Journal of the Audio Engineering Society*, *53*(7/8), 620–631.

Hall III, J. W. (2007). *New Handbook of Auditory Evoked Responses*. Pearson.

Hall III, J. W., Grose, J. H., Buss, E., & Dev, M. B. (2002). Spondee Recognition in a Two-Talker Masker and a Speech-Shaped Noise Masker in Adults and Children. *Ear and Hearing*, *23*(2), 159–165. https://doi.org/10.1097/00003446-200204000-00008

Hamacher, V., Chalupper, J., Eggers, J., Fischer, E., Kornagel, U., Puder, H., & Rass, U. (2005). Signal Processing in High-End Hearing Aids: State of the Art, Challenges, and Future Trends. *EURASIP Journal on Advances in Signal Processing*, *2005*(18), 1–15. https://doi.org/10.1155/ASP.2005.2915

Hardin, R. H., & Sloane, N. J. (1996). McLaren's improved snub cube and other new spherical designs in three dimensions. *Discrete & Computational Geometry*, *15*(4), 429–441. https://doi.org/10.1007/BF02711518

Hartley, R. V. L., & Fry, T. C. (1921). The Binaural Location of Pure Tones. *Phys. Rev.*, *18*, 431–442. https://doi.org/10.1103/PhysRev.18.431

Hartmann, W., & Macaulay, E. (2014). Anatomical limits on interaural time differences: An ecological perspective. *Frontiers in Neuroscience*, *8*, 34. https://doi.org/10.3389/fnins.2014.00034

Hawley, M. L., Litovsky, R. Y., & Culling, J. F. (2004). The benefit of binaural hearing in a cocktail party: Effect of location and type of interferer. *The Journal of the Acoustical Society of America*, 115(2), 833–843. https://doi.org/10.1121/1.1639908

Healy, E. W., Yoho, S. E., Wang, Y., & Wang, D. (2013). An algorithm to improve speech recognition in noise for hearing-impaired listeners. *The Journal of the Acoustical Society of America*, 134(4), 3029–3038. https://doi.org/10.1121/1.4820893

Hebrank, J., & Wright, D. (1974). Spectral cues used in the localization of sound sources on the median plane. *The Journal of the Acoustical Society of America*, 56(6), 1829–1834. https://doi.org/10.1121/1.1903520

Hellbrück, J., & Liebl, A. (2008). Noise effects on cognitive performance. In *Kuwano, sonoko (hrsg.): Recent topics in environmental psychoacoustics* (pp. 153–184). Osaka University Press.

Hendrikse, M. M. E., Llorach, G., Hohmann, V., & Grimm, G. (2019). Movement and Gaze Behavior in Virtual Audiovisual Listening Environments Resembling Everyday Life [PMID: 32516060]. *Trends in Hearing*, 23, 2331216519872362. https://doi.org/10.1177/2331216519872362

Henshaw, H., & Ferguson, M. A. (2013). Efficacy of Individual Computer-Based Auditory Training for People with Hearing Loss: A Systematic Review of the Evidence. *PloS one*, 8(5), e62836. https://doi.org/10.1371/journal.pone.0062836

Herzke, T., Kayser, H., Loshaj, F., Grimm, G., & Hohmann, V. (2017). Open signal processing software platform for hearing aid research (openmha). *Proceedings of the Linux Audio Conference, Saint-Etienne, France*, 35–42.

Hochmair-Desoyer et al. (1997). The HSM sentence test as a tool for evaluating the speech understanding in noise of cochlear implant users. *The American Journal of Otology*, 18(Suppl 6), S83.

Hogan, A., Shipley, M., Strazdins, L., Purcell, A., & Baker, E. (2011). Communication and behavioural disorders among children with hearing loss increases risk of mental health disorders. *Australian and New Zealand journal of public health*, 35(4), 377–383.

Hohmann, V. (2002). Frequency analysis and synthesis using a Gammatone filterbank. *Acta Acustica united with Acustica*, 88(3), 433–442.

Holm, S. (1979). A Simple Sequentially Rejective Multiple Test Procedure. *Scandinavian Journal of Statistics*, 6(2), 65–70.

Hougaard, S., & Ruf, S. (2011). EuroTrak I: A consumer survey about hearing aids in Germany, France, and the UK. *Hearing Review*, 18(2), 12–28.

Houtgast, T., & Steeneken, H. J. M. (1985). A review of the MTF concept in room acoustics and its use for estimating speech intelligibility in auditoria. *The Journal of the Acoustical Society of America*, 77(3), 1069–1077. https://doi.org/10.1121/1.392224

Houtgast, T., & Festen, J. M. (2008). On the auditory and cognitive functions that may explain an individual's elevation of the speech reception threshold in noise. *International Journal of Audiology*, 47(6), 287–295. https://doi.org/10.1080/14992020802127109

Hygge, S., Evans, G. W., & Bullinger, M. (2002). A prospective study of some effects of aircraft noise on cognitive performance in schoolchildren. *Psychological science, 13*(5), 469–474. https://doi.org/10.1111/1467-9280.00483

ICD-10-GM. (2020, May 26). International Statistical Classification Of Diseases And Related Health Problems, 10th revision, German Modification (ICD-10-GM), Federal Institute for Drugs and Medical Devices, Germany [https://www.dimdi.de/dynamic/en/classifications/icd/icd-10-gm/].

IEC 60318-5:2006. (2006, August). *Electroacoustics – Simulators of human head and ear - Part 5: 2 cm3 coupler for the measurement of hearing aids and earphones coupled to the ear by means of ear inserts* (Norm IEC 60318-5:2006). International Electrotechnical Commission. Geneva, Switzerland.

Iglehart, F. (2020). Speech Perception in Classroom Acoustics by Children With Hearing Loss and Wearing Hearing Aids. *American Journal of Audiology, 29*(1), 6–17. https://doi.org/10.1044/2019_AJA-19-0010

Iida, K., Itoh, M., Itagaki, A., & Morimoto, M. (2007). Median plane localization using a parametric model of the head-related transfer function based on spectral cues. *Applied Acoustics, 68*(8), 835–850. https://doi.org/10.1016/j.apacoust.2006.07.016

ISO 226:2003. (2003). *Acoustics – Normal equal-loudness-level contours* (Standard). International Organization for Standardization. Geneva, Switzerland.

ISO 3382-1. (2009). *Acoustics – Measurement of room acoustic parameters – Part 1: Performance spaces* (Norm ISO 3382-1:2009-06). International Organization for Standardization. Geneva, Switzerland.

ISO 3382-2. (2008). *Acoustics – Measurement of room acoustic parameters - Part 2: Reverberation time in ordinary rooms* (Norm ISO 3382-2:2008-06). International Organization for Standardization. Geneva, Switzerland.

ISO 8253-1. (2010). *Acoustics – Audiometric test methods – Part 1: Pure-tone air and bone conduction audiometry* (Norm ISO 8253-1:2010). International Organization for Standardization. Geneva, Switzerland.

ISO 8253-2. (2009). *Acoustics – Audiometric test methods – Part 2: Sound field audiometry with pure-tone and narrow-band test signals* (Norm ISO 8253-2:2009). International Organization for Standardization. Geneva, Switzerland.

ISO 8253-3. (2012). *Acoustics – Audiometric test methods – Part 3: Speech audiometry (ISO 8253-3:2012)* (Norm ISO 8253-3:2012). International Organization for Standardization. Geneva, Switzerland.

Jones, G. (2021). Psychometriccurvefitting [Available at https://github.com/garethjns/PsychometricCurveFitting, accessed on 2021-07-23].

Jot, J.-M., Wardle, S., & Larcher, V. Approaches to Binaural Synthesis. In: In *Audio engineering society convention 105*. 1998, September. http://www.aes.org/e-lib/browse.cfm?elib=8319

Jungmann, J. O., Mazur, R., Kallinger, M., Mei, T., & Mertins, A. (2012). Combined Acoustic MIMO Channel Crosstalk Cancellation and Room Impulse Response Reshaping. *IEEE Transactions on Audio, Speech, and Language Processing, 20*(6), 1829–1842.

Kates, J. M. (2001). Room reverberation effects in hearing aid feedback cancellation. *The Journal of the Acoustical Society of America, 109*(1), 367–378. https://doi.org/10.1121/1.1332379

Kates, J. M. (2003). Adaptive Feedback Cancellation in Hearing Aids. In *Adaptive signal processing* (pp. 23–57). Springer.

Kates, J. M. (2008). *Digital Hearing Aids*. Plural publishing.

Kates, J. M. (2010). Understanding compression: Modeling the effects of dynamic-range compression in hearing aids. *International Journal of Audiology, 49*(6), 395–409. https://doi.org/10.3109/14992020903426256

Katz, B. F. (2001). Boundary element method calculation of individual head-related transfer function. I. Rigid model calculation. *The Journal of the Acoustical Society of America, 110*(5), 2440–2448. https://doi.org/10.1121/1.1412440

Katz, B. F., & Noisternig, M. (2014). A comparative study of interaural time delay estimation methods. *The Journal of the Acoustical Society of America, 135*(6), 3530–3540. https://doi.org/10.1121/1.4875714

Kayser, H., Ewert, S. D., Anemüller, J., Rohdenburg, T., Hohmann, V., & Kollmeier, B. (2009). Database of Multichannel In-ear and Behind-the-ear Head-related and Binaural Room Impulse Responses. *EURASIP J. Adv. Signal Process, 2009*, 6:1–6:10. https://doi.org/10.1155/2009/298605

Kearney, G., Gorzel, M., Rice, H., & Boland, F. (2012). Distance Perception in Interactive Virtual Acoustic Environments using First and Higher Order Ambisonic Sound Fields. *Acta Acustica united with Acustica, 98*(1), 61–71. https://doi.org/doi:10.3813/AAA.918492

Keidser, G., & Convery, E. (2016). Self-Fitting Hearing Aids: Status Quo and Future Predictions [PMID: 27072929]. *Trends in Hearing, 20*, 2331216516643284. https://doi.org/10.1177/2331216516643284

Keidser, G., Dillon, H., Flax, M., Ching, T., & Brewer, S. (2011). The NAL-NL2 prescription procedure. *Audiology research, 1*(1), 88–90.

Keidser, G., Rohrseitz, K., Dillon, H., Hamacher, V., Carter, L., Rass, U., & Convery, E. (2006). The effect of multi-channel wide dynamic range compression, noise reduction, and the directional microphone on horizontal localization performance in hearing aid wearers. *International Journal of Audiology, 45*(10), 563–579.

Keiner, J., Kunis, S., & Potts, D. (2007). Efficient reconstruction of functions on the sphere from scattered data. *Journal of Fourier Analysis and Applications, 13*(4), 435–458.

Kidd, G., Mason, C. R., Brughera, A., & Hartmann, W. M. (2005). The role of reverberation in release from masking due to spatial separation of sources for speech identification. *Acta acustica united with acustica, 91*(3), 526–536.

Killion, M. C., & Fikret-Pasa, S. (1993). The 3 types of sensorineural hearing loss: Loudness and intelligibility considerations. *Hearing journal, 46*, 31–31.

Kistler, D. J., & Wightman, F. L. (1992). A model of head-related transfer functions based on principal components analysis and minimum-phase reconstruction. *The*

Journal of the Acoustical Society of America, 91(3), 1637–1647. https://doi.org/10.1121/1.402444

Klatte, M., Hellbrück, J., Seidel, J., & Leistner, P. (2010a). Effects of Classroom Acoustics on Performance and Well-Being in Elementary School Children: A Field Study. *Environment and Behavior, 42*(5), 659–692. https://doi.org/10.1177/0013916509336813

Klatte, M., Lachmann, T., & Meis, M. (2010). Effects of noise and reverberation on speech perception and listening comprehension of children and adults in a classroom-like setting. *Noise and Health, 12*(49), 270–282. https://doi.org/10.4103/1463-1741.70506

Klatte, M., Lachmann, T., Schlittmeier, S., & Hellbrück, J. (2010b). The irrelevant sound effect in short-term memory: Is there developmental change? *European Journal of Cognitive Psychology, 22*(8), 1168–1191. https://doi.org/10.1080/09541440903378250

Klein, J. C. (2020). *Directional room impulse response measurement* (Doctoral dissertation). Institute of Technical Acoustics, RWTH Aachen University. Aachen, Germany, Logos Verlag. https://doi.org/10.18154/RWTH-2020-07829

Kock, W. (1950). Binaural localization and masking. *The Journal of the Acoustical Society of America, 22*(6), 801–804.

Kohnen, M., Stienen, J., Aspöck, L., & Vorländer, M. (2016). Performance Evaluation of a Dynamic Crosstalk-Cancellation System with Compensation of Early Reflections. *Audio Engineering Society Conference: 2016 AES International Conference on Sound Field Control*, 1–8.

Kolarik, A. J., Moore, B. C., Zahorik, P., Cirstea, S., & Pardhan, S. (2015). Auditory distance perception in humans: a review of cues, development, neuronal bases, and effects of sensory loss. *Attention, Perception, & Psychophysics*, 1–23. https://doi.org/10.3758/s13414-015-1015-1

Kotel'nikov, V. A. (1933). On the carrying capacity of the "either" and wire in telecommunications. *Material for the First All-Union Conference on Questions of Communication (Russian), Izd. Red. Upr. Svyzai RKKA, Moscow, 1933.*

Krokstad, A., Strom, S., & Sørsdal, S. (1968). Calculating the acoustical room response by the use of a ray tracing technique. *Journal of Sound and Vibration, 8*(1), 118–125. https://doi.org/10.1016/0022-460X(68)90198-3

Kuk, F. K. (1996). Theoretical and Practical Considerations in Compression Hearing Aids [PMID: 25425854]. *Trends in Amplification, 1*(1), 5–39. https://doi.org/10.1177/108471389600100102

Kuttruff, H. (2016). *Room Acoustics*. CRC Press.

Kuznetsova, A., Brockhoff, P. B., & Christensen, R. H. B. (2017). lmerTest Package: Tests in Linear Mixed Effects Models. *Journal of Statistical Software, 82*(13), 1–26. https://doi.org/10.18637/jss.v082.i13

Laitinen, M.-V., Pihlajamäki, T., Lösler, S., & Pulkki, V. Influence of Resolution of Head Tracking in Synthesis of Binaural Audio. In: In *Audio engineering society convention 132*. Budapest, Hungary: Audio Engineering Society, 2012, April, 1–8.

Lee, S.-l., Kim, L.-H., & Sung, K.-M. Head Related Transfer Function Refinement Using Directional Weighting Function. In: In *Audio engineering society convention 115*. 2013, October. http://www.aes.org/e-lib/browse.cfm?elib=12428

Leek, M. R. (2001). Adaptive procedures in psychophysical research. *Perception & psychophysics, 63*(8), 1279–1292.

Lentz, T. (2008). *Binaural technology for virtual reality* (Doctoral dissertation). Institute of Technical Acoustics, RWTH Aachen University, Germany. Aachen, Germany, Logos Verlag Berlin GmbH.

Lentz, T., Schröder, D., Vorländer, M., & Assenmacher, I. (2007). Virtual Reality System with Integrated Sound Field Simulation and Reproduction. *EURASIP J. Appl. Signal Process., 2007*(1), 187–187. https://doi.org/10.1155/2007/70540

Levitt, H. (1971). Transformed up-down methods in psychoacoustics. *The Journal of the Acoustical society of America, 49*(2B), 467–477.

Levy-Shiff, R., & Hoffman, M. A. (1985). Social behaviour of hearing-impaired and normally-hearing preschoolers. *British Journal of Educational Psychology, 55*(2), 111–118. https://doi.org/https://doi.org/10.1111/j.2044-8279.1985.tb02615.x

Lidén, G., Nordlund, B., & Hawkins, J. E. (1964). Significance of the Stapedius Reflex for the Understanding of Speech. *Acta Oto-Laryngologica, 57*(sup188), 275–279. https://doi.org/10.3109/00016486409134576

Lidén, G., Peterson, J. L., & Björkman, G. (1970). Tympanometry. *Archives of Oto-laryngology, 92*(3), 248–257.

Lindau, A. (2009). The Perception of System Latency in Dynamic Binaural Synthesis. *Fortschritte der Akustik: Tagungsband der 35. DAGA*, 1063–1066.

Lindau, A., Erbes, V., Lepa, S., Maempel, H.-J., Brinkmann, F., & Weinzierl, S. (2014). A Spatial Audio Quality Inventory (SAQI). *Acta Acustica united with Acustica, 100*(5), 984–994. https://doi.org/10.3813/AAA.918778

Lindau, A., Kosanke, L., & Weinzierl, S. (2012). Perceptual Evaluation of Model- and Signal-Based Predictors of the Mixing Time in Binaural Room Impulse Responses. *Journal of the Audio Engineering Society, 60*(11), 887–898.

Lindau, A., Maempel, H.-J., & Weinzierl, S. (2008). Minimum BRIR grid resolution for dynamic binaural synthesis. *The Journal of the Acoustical Society of America, 123*(5), 3498–3498. https://doi.org/10.1121/1.2934364

Lindau, A., & Weinzierl, S. (2012). Assessing the plausibility of virtual acoustic environments. *Acta Acustica united with Acustica, 98*(5), 804–810.

Litovsky, R. Y. (2005). Speech intelligibility and spatial release from masking in young children. *The Journal of the Acoustical Society of America, 117*(5), 3091–3099. https://doi.org/10.1121/1.1873913

Litovsky, R. Y. (2012). Spatial release from masking. *Acoustics today, 8*(2), 18–25.

Litovsky, R. Y., Colburn, H. S., Yost, W. A., & Guzman, S. J. (1999). The precedence effect. *The Journal of the Acoustical Society of America, 106*(4), 1633–1654.

Lopez, J. J., Gutierrez, P., Cobos, M., & Aguilera, E. (2014). Sound distance perception comparison between Wave Field Synthesis and Vector Base Amplitude Panning. *2014*

6th International Symposium on Communications, Control and Signal Processing (ISCCSP), 165–168. https://doi.org/10.1109/ISCCSP.2014.6877841

Lüdecke, D. (2021). *sjPlot: Data Visualization for Statistics in Social Science* [Available at https://CRAN.R-project.org/package=sjPlot (Version 2.8.7)].

Macpherson, E. A., & Middlebrooks, J. C. (2002). Listener weighting of cues for lateral angle: The duplex theory of sound localization revisited. *The Journal of the Acoustical Society of America, 111*(5), 2219–2236. https://doi.org/10.1121/1.1471898

Majdak, P., Balazs, P., & Laback, B. (2007). Multiple Exponential Sweep Method for Fast Measurement of Head-Related Transfer Functions. *Journal of the Audio Engineering Society, 55*(7/8), 623–637.

Majdak, P., Iwaya, Y., Carpentier, T., Nicol, R., Parmentier, M., Roginska, A., Suzuki, Y., Watanabe, K., Wierstorf, H., Ziegelwanger, H., et al. (2013). Spatially oriented format for acoustics: A data exchange format representing head-related transfer functions. *Audio Engineering Society Convention 134*, 1–11.

Majdak, P., Masiero, B., & Fels, J. (2013). Sound localization in individualized and non-individualized crosstalk cancellation systems. *The Journal of the Acoustical Society of America, 133*(4), 2055–2068. https://doi.org/10.1121/1.4792355

Makita, Y. (1962). On the directional localization of sound in the stereophonic sound field. *EBU review, 73*(Part A), 1536–1539.

Makous, J. C., & Middlebrooks, J. C. (1990). Two-dimensional sound localization by human listeners. *The journal of the Acoustical Society of America, 87*(5), 2188–2200.

Malham, D. G., & Myatt, A. (1995). 3-D Sound Spatialization using Ambisonic Techniques. *Computer Music Journal, 19*(4), 58–70.

Markley, F. L., & Crassidis, J. L. (2014). *Fundamentals of Spacecraft Attitude Determination and Control.* Springer.

Marrone, N., Mason, C. R., & Kidd, G. (2008). The effects of hearing loss and age on the benefit of spatial separation between multiple talkers in reverberant rooms. *The Journal of the Acoustical Society of America, 124*(5), 3064–3075. https://doi.org/10.1121/1.2980441

Marshall, L. G. (1994). An acoustics measurement program for evaluating auditoriums based on the early/late sound energy ratio. *The Journal of the Acoustical Society of America, 96*(4), 2251–2261. https://doi.org/10.1121/1.410097

Masiero, B., & Fels, J. (2011). Perceptually Robust Headphone Equalization for Binaural Reproduction. *Audio Engineering Society Convention 130*, 1–7.

McAnally, K. I., & Martin, R. L. (2014). Sound localization with head movement: Implications for 3-d audio displays. *Frontiers in neuroscience, 8*.

McKeag, A., & McGrath, D. S. Sound Field Format to Binaural Decoder with Head Tracking. In: In *Audio engineering society convention 6r.* 1996, August. http://www.aes.org/e-lib/browse.cfm?elib=7477

Mehra, R., Rungta, A., Golas, A., Lin, M., & Manocha, D. (2015). WAVE: Interactive Wave-based Sound Propagation for Virtual Environments. *IEEE Transactions on*

Visualization and Computer Graphics, 21(4), 434–442. https://doi.org/10.1109/
TVCG.2015.2391858

Mener, D. J., Betz, J., Genther, D. J., Chen, D., & Lin, F. R. (2013). Hearing Loss
and Depression in Older Adults. *Journal of the American Geriatrics Society, 61*(9),
1627–1629. https://doi.org/https://doi.org/10.1111/jgs.12429

Mershon, D. H., & King, L. E. (1975). Intensity and reverberation as factors in the
auditory perception of egocentric distance. *Perception & Psychophysics, 18*(6), 409–
415. https://doi.org/https://doi.org/10.3758/BF03204113

Messaoud-Galusi, S., Hazan, V., & Rosen, S. (2011). Investigating Speech Perception in
Children With Dyslexia: Is There Evidence of a Consistent Deficit in Individuals?
Journal of Speech, Language, and Hearing Research, 54(6), 1682–1701. https://doi.
org/10.1044/1092-4388(2011/09-0261)

Meyer, J., & Elko, G. A Qualitative Analysis of Frequency Dependencies in Ambisonics
Decoding Related to Spherical Microphone Array Recording. In: In *Audio engineering society conference: 2016 aes international conference on sound field control.*
2016, July. http://www.aes.org/e-lib/browse.cfm?elib=18316

Mick, P., Kawachi, I., & Lin, F. R. (2014). The Association between Hearing Loss and
Social Isolation in Older Adults [PMID: 24384545]. *Otolaryngology–Head and Neck
Surgery, 150*(3), 378–384. https://doi.org/10.1177/0194599813518021

Middlebrooks, J. C. (1999). Individual differences in external-ear transfer functions
reduced by scaling in frequency. *The Journal of the Acoustical Society of America,
106*(3), 1480–1492. https://doi.org/10.1121/1.427176

Middlebrooks, J. C., & Green, D. M. (1991). Sound localization by human listeners.
Annual review of psychology, 42(1), 135–159. https://doi.org/10.1146/annurev.ps.
42.020191.001031

Mills, A. W. (1958). On the Minimum Audible Angle. *The Journal of the Acoustical
Society of America, 30*(4), 237–246. https://doi.org/10.1121/1.1909553

Mills, A. W. (1972). *Auditory Localization* (J. Tobias, Ed.). Academic Press.

Mills, M. (2011). Hearing Aids and the History of Electronics Miniaturization. *IEEE
Annals of the History of Computing, 33*(2), 24–45. https://doi.org/10.1109/MAHC.
2011.43

Minnaar, P., Favrot, S., & Buchholz, J. M. (2010). Improving hearing aids through
listening tests in a virtual sound environment. *The Hearing Journal, 63*(10), 40–42.
https://doi.org/10.1097/01.HJ.0000389926.64797.3e

Misurelli, S. M., & Litovsky, R. Y. (2015). Spatial release from masking in children
with bilateral cochlear implants and with normal hearing: Effect of target-interferer
similarity. *The Journal of the Acoustical Society of America, 138*(1), 319–331. https:
//doi.org/10.1121/1.4922777

Moeck, T., Bonneel, N., Tsingos, N., Drettakis, G., Viaud-Delmon, I., & Alloza, D.
(2007). Progressive Perceptual Audio Rendering of Complex Scenes. *Proceedings
of the 2007 Symposium on Interactive 3D Graphics and Games*, 189–196. https:
//doi.org/10.1145/1230100.1230133

Møller, H. (1992). Fundamentals of binaural technology. *Applied acoustics*, *36*(3-4), 171–218.

Møller, H., Jensen, C. B., Hammershøi, D., & Sørensen, M. F. (1995a). Design Criteria for Headphones. *Journal of the Audio Engineering Society*, *43*(4), 218–232.

Møller, H., Sørensen, M. F., Hammershøi, D., & Jensen, C. B. (1995b). Head-Related Transfer Functions of Human Subjects. *Journal of the Audio Engineering Society*, *43*(5), 300–321.

Moore, B. C. (2012). *An Introduction to the Psychology of Hearing*. Brill.

Moore, B. C., Alcantara, J. I., Stone, M. A., & Glasberg, B. R. (1999). Use of a loudness model for hearing aid fitting: II. Hearing aids with multi-channel compression [PMID: 10439142]. *British Journal of Audiology*, *33*(3), 157–170. https://doi.org/10.3109/03005369909090095

Moorman, S. M., Greenfield, E. A., & Lee, C. S. H. (2021). Perceived Hearing Loss, Social Disengagement, and Declines in Memory [PMID: 32108530]. *Journal of Applied Gerontology*, *40*(6), 679–683. https://doi.org/10.1177/0733464820909244

Mueller, M. F., Kegel, A., Schimmel, S. M., Dillier, N., & Hofbauer, M. (2012). Localization of virtual sound sources with bilateral hearing aids in realistic acoustical scenes. *The Journal of the Acoustical Society of America*, *131*(6), 4732–4742. https://doi.org/10.1121/1.4705292

Müller, S., & Massarani, P. (2001). Transfer-Function Measurement with Sweeps. *Journal of the Audio Engineering Society*, *49*(6), 443–471.

Musicant, A. D., & Butler, R. A. (1985). Influence of monaural spectral cues on binaural localization. *The Journal of the Acoustical Society of America*, *77*(1), 202–208.

Nash, J. C. (2014). On Best Practice Optimization Methods in R. *Journal of Statistical Software*, *60*(2), 1–14.

Nash, J. C., & Varadhan, R. (2011). Unifying Optimization Algorithms to Aid Software System Users: optimx for R. *Journal of Statistical Software*, *43*(9), 1–14.

Nelson, D. I., Nelson, R. Y., Concha-Barrientos, M., & Fingerhut, M. (2005). The global burden of occupational noise-induced hearing loss. *American Journal of Industrial Medicine*, *48*(6), 446–458. https://doi.org/10.1002/ajim.20223

Nelson, J. J., & Chen, K. (2004). The Relationship of Tinnitus, Hyperacusis, and Hearing Loss [PMID: 15372918]. *Ear, Nose & Throat Journal*, *83*(7), 472–476. https://doi.org/10.1177/014556130408300713

Nelson, P. B., Perry, T. T., Gregan, M., & VanTasell, D. (2018). Self-Adjusted Amplification Parameters Produce Large Between-Subject Variability and Preserve Speech Intelligibility [PMID: 30191767]. *Trends in Hearing*, *22*. https://doi.org/10.1177/2331216518798264

Nelson, P. A., & Elliott, S. J. (1995). *Active Control of Sound* (3rd ed.). Academic press.

Neuman, A. C., & Hochberg, I. (1983). Children's perception of speech in reverberation. *The Journal of the Acoustical Society of America*, *73*(6), 2145–2149. https://doi.org/10.1121/1.389538

Nicol, R., Gros, L., Colomes, C., Noisternig, M., Warusfel, O., Bahu, H., Katz, B. F., & Simon, L. S. (2014). A Roadmap for Assessing the Quality of Experience of 3D Audio Binaural Rendering. *Proceedings of the EAA Joint Symposium on Auralization and Ambisonics 2014*, 100–106. https://doi.org/10.14279/depositonce-4103

Niemeyer, W., & Sesterhenn, G. (1974). Calculating the Hearing Threshold from the Stapedius Reflex Threshold for Different Sound Stimuli. *Audiology*, *13*(5), 421–427. https://doi.org/10.3109/00206097409071701

Nikles, J.-M., & Tschopp, K. (1996). Audiological applications of the Basle Sentence Intelligibility Test. *Audiological Acoustics*, *35*, 70–75.

Nittrouer, S., Caldwell-Tarr, A., Tarr, E., Lowenstein, J. H., Rice, C., & Moberly, A. C. (2013). Improving speech-in-noise recognition for children with hearing loss: Potential effects of language abilities, binaural summation, and head shadow. *International Journal of Audiology*, *52*(8), 513–525. https://doi.org/10.3109/14992027.2013.792957

Noisternig, M., Katz, B. F. G., Siltanen, S., & Savioja, L. (2008). Framework for Real-Time Auralization in Architectural Acoustics. *Acta Acustica united with Acustica*, *94*(6), 1000–1015. https://doi.org/10.3813/AAA.918116

Oberem, J. (2020). *Examining auditory selective attention: From dichotic towards realistic environments* (Doctoral dissertation). RWTH Aachen University. Aachen, Germany, Logos Verlag GmbH. https://doi.org/10.18154/RWTH-2020-04889

Oberem, J., Koch, I., & Fels, J. (2017). Intentional switching in auditory selective attention: Exploring age-related effects in a spatial setup requiring speech perception. *Acta Psychologica*, *177*, 36–43. https://doi.org/https://doi.org/10.1016/j.actpsy.2017.04.008

Oberem, J., Lawo, V., Koch, I., & Fels, J. (2014). Intentional Switching in Auditory Selective Attention: Exploring Different Binaural Reproduction Methods in an Anechoic Chamber. *Acta Acustica united with Acustica*, *100*(6), 1139–1148. https://doi.org/doi:10.3813/AAA.918793

Oberem, J., Masiero, B., & Fels, J. (2016). Experiments on authenticity and plausibility of binaural reproduction via headphones employing different recording methods. *Applied Acoustics*, *114*, 71–78. https://doi.org/https://doi.org/10.1016/j.apacoust.2016.07.009

Oberem, J., Richter, J.-G., Setzer, D., Seibold, J., Koch, I., & Fels, J. (2020). Experiments on localization accuracy with non-individual and individual hrtfs comparing static and dynamic reproduction methods. *bioRxiv*. https://doi.org/10.1101/2020.03.31.011650

OpenDAFF. (2018, January). An open source file format for directional audio content, Institute of Technical Acoustics, RWTH Aachen University [Available at http://www.opendaff.org/, accessed on 2021-06-23].

Oppenheim, A. V., Schafer, R. W., & Buck, J. R. (1999). *Discrete-Time Signal Processing*. Prentice Hall.

Oreinos, C., & Buchholz, J. M. (2015). Objective analysis of ambisonics for hearing aid applications: Effect of listener's head, room reverberation, and directional microphones. *The Journal of the Acoustical Society of America*, *137*(6), 3447–3465. https://doi.org/10.1121/1.4919330

Oreinos, C., & Buchholz, J. M. (2016). Evaluation of Loudspeaker-Based Virtual Sound Environments for Testing Directional Hearing Aids. *Journal of the American Academy of Audiology*, *27*(7), 541–556. https://doi.org/10.3766/jaaa.15094

Palmer, C. V., & Lindley, G. (2002). Overview and Rationale for Prescriptive Formulas for Linear and Nonlinear Hearing Aids. *Strategies for selecting and verifying hearing aid fittings (2nd ed.)*. New York, NY: Thieme Medical Publishers.

Parodi, Y. L., & Rubak, P. (2010). Objective evaluation of the sweet spot size in spatial sound reproduction using elevated loudspeakers. *The Journal of the Acoustical Society of America*, *128*(3), 1045–1055. https://doi.org/10.1121/1.3467763

Parodi, Y. L., & Rubak, P. (2011). A Subjective Evaluation of the Minimum Channel Separation for Reproducing Binaural Signals over Loudspeakers. *Journal of the Audio Engineering Society*, *59*(7/8), 487–497.

Patel, S. R., Bouldin, E., Tey, C. S., Govil, N., & Alfonso, K. P. (2021). Social Isolation and Loneliness in the Hearing-Impaired Pediatric Population: A Scoping Review. *The Laryngoscope*, *131*(8), 1869–1875. https://doi.org/https://doi.org/10.1002/lary.29312

Pearson, E. S. (1931). The Analysis of Variance in Cases of Non-Normal Variation. *Biometrika*, *23*(1/2), 114–133. https://doi.org/10.1080/00949650213745

Pelzer, S., Aretz, M., & Vorländer, M. (2011). Quality assessment of room acoustic simulation tools by comparing binaural measurements and simulations in an optimized test scenario. *Proceedings of Forum Acusticum 2011 : 27 June - 01 July, Aalborg, Denmark / ed. by Danish Acoustical Society (DAS) on behalf of European Acoustics Association (EAA)*, 1529–1534.

Pelzer, S., Aspöck, L., Schröder, D., & Vorländer, M. (2014). Interactive Real-Time Simulation and Auralization for Modifiable Rooms. *Building Acoustics*, *21*(1), 65–73. https://doi.org/10.1260/1351-010X.21.1.65

Pelzer, S., Masiero, B., & Vorländer, M. (2014). 3D reproduction of room auralizations by combining intensity panning, crosstalk cancellation and Ambisonics. *Proceedings of the EAA Joint Symposium on Auralization and Ambisonics 2014*, 182–188. https://doi.org/10.14279/depositonce-33

Pentland, A. (1980). Maximum likelihood estimation: The best PEST. *Perception & psychophysics*.

Perrott, D. R., & Pacheco, S. (1989). Minimum audible angle thresholds for broadband noise as a function of the delay between the onset of the lead and lag signals. *The Journal of the Acoustical Society of America*, *85*(6), 2669–2672. https://doi.org/10.1121/1.397764

Peyre, G. (2021, May 19). Toolbox Graph [Available at https://www.mathworks.com/matlabcentral/fileexchange/5355-toolbox-graph, accessed on 2021-08-25].

Picou, E. M., & Ricketts, T. A. (2019). An Evaluation of Hearing Aid Beamforming Microphone Arrays in a Noisy Laboratory Setting. *Journal of the American Academy of Audiology*, *30*(2), 131–144.

Plogsties, J., Minnaar, P., Olesen, S. K., Christensen, F., & Møller, H. Audibility of All-Pass Components in Head-Related Transfer Functions. In: In *Audio engineering society convention 108*. 2000, February. http://www.aes.org/e-lib/browse.cfm? elib=9206

Plomp, R. (1976). Binaural and monaural speech intelligibility of connected discourse in reverberation as a function of azimuth of a single competing sound source (speech or noise). *Acta Acustica united with Acustica, 34*(4), 200–211.

Plomp, R. (1978). Auditory handicap of hearing impairment and the limited benefit of hearing aids. *The Journal of the Acoustical Society of America, 63*(2), 533–549.

Poletti, M. A. (2005). Three-Dimensional Surround Sound Systems Based on Spherical Harmonics. *Journal of the Audio Engineering Society, 53*(11), 1004–1025.

Politis, A. (2021a). *Matlab/Octave library "Higher-Order-Ambisonics"* [Available at https://github.com/polarch/Higher-Order-Ambisonics.git]. GitHub.

Politis, A. (2021b). *Matlab/Octave library "Vector-Base-Amplitude-Panning"* [Available at https://github.com/polarch/Vector-Base-Amplitude-Panning.git]. GitHub.

Prabhu, K. M. (2014). *Window Functions and Their Applications in Signal Processing*. Taylor & Francis.

Preparata, F. P., & Shamos, M. I. (1985). Convex Hulls: Basic Algorithms. In *Computational geometry* (pp. 95–149). Springer.

Probst, R., Lonsbury-Martin, B. L., & Martin, G. K. (1991). A review of otoacoustic emissions. *The Journal of the Acoustical Society of America, 89*(5), 2027–2067.

Pulkki, V. (1997). Virtual Sound Source Positioning Using Vector Base Amplitude Panning. *Journal of the Audio Engineering Society, 45*(6), 456–466.

Pulkki, V. (1999). Uniform spreading of amplitude panned virtual sources. *Applications of Signal Processing to Audio and Acoustics, 1999 IEEE Workshop on*, 187–190. https://doi.org/10.1109/ASPAA.1999.810881

Pulkki, V. (2001). *Spatial sound generation and perception by amplitude panning techniques* (Doctoral dissertation). Helsinki University of Technology Laboratory of Acoustics and Audio Signal Processing.

Pulkki, V., Karjalainen, M., & Välimäki, V. Localization, Coloration, and Enhancement of Amplitude-Panned Virtual Sources. In: In *Audio engineering society conference: 16th international conference: Spatial sound reproduction*. 1999, March. http://www.aes.org/e-lib/browse.cfm?elib=8030

R Core Team. (2021). *R: A Language and Environment for Statistical Computing* [Available at https://www.R-project.org/ (R version 4.0.4, 2021-02-15)]. R Foundation for Statistical Computing. Vienna, Austria.

Raake, A., Wierstorf, H., & Blauert, J. (2014). A case for TWO! EARS in audio quality assessment. *Forum Acusticum*, 41.

Rakerd, B., & Hartmann, W. M. (2010). Localization of sound in rooms. V. Binaural coherence and human sensitivity to interaural time differences in noise. *The Journal of the Acoustical Society of America, 128*(5), 3052–3063. https://doi.org/10.1121/1.3493447

Rana, B., & Buchholz, J. M. (2018). Effect of audibility on better-ear glimpsing as a function of frequency in normal-hearing and hearing-impaired listeners. *The Journal of the Acoustical Society of America, 143*(4), 2195–2206. https://doi.org/10.1121/1.5031007

Raykar, V. C., Duraiswami, R., & Yegnanarayana, B. (2005). Extracting the frequencies of the pinna spectral notches in measured head related impulse responses. *The Journal of the Acoustical Society of America, 118*(1), 364–374. https://doi.org/10.1121/1.4785467

Rayleigh, L. (1907). XII. On our perception of sound direction. *The London, Edinburgh, and Dublin Philosophical Magazine and Journal of Science, 13*(74), 214–232. https://doi.org/10.1080/14786440709463595

Reichardt, W., Alim, O., & Schmidt, W. (1974). Abhängigkeit der Grenzen zwischen brauchbarer und unbrauchbarer Durchsichtigkeit von der Art des Musikmotives, der Nachhallzeit und der Nachhalleinsatzzeit (Dependency of the limits between usable and unusable clarity on the type of musical motif, the reverberation time and the reverberation onset time). *Applied Acoustics, 7*(4), 243–264. https://doi.org/https://doi.org/10.1016/0003-682X(74)90033-4

Richardson, J. T. E., Long, G. L., & Woodley, A. (2004). Students with an Undisclosed Hearing Loss: A Challenge for Academic Access, Progress, and Success? *The Journal of Deaf Studies and Deaf Education, 9*(4), 427–441. https://doi.org/10.1093/deafed/enh044

Richter, J.-G. (2019). *Fast measurement of individual head-related transfer functions* (Doctoral dissertation). RWTH Aachen University. Aachen, Germany, Logos Verlag Berlin GmbH. https://doi.org/10.18154/RWTH-2019-04006

Richter, J.-G., & Fels, J. (2016). Evaluation of Localization Accuracy of Static Sources Using HRTFs from a Fast Measurement System. *Acta Acustica united with Acustica, 102*(4), 763–771. https://doi.org/10.3813/AAA.918992

Richter, J.-G., & Fels, J. (2019). On the Influence of Continuous Subject Rotation During High-Resolution Head-Related Transfer Function Measurements. *IEEE/ACM Transactions on Audio, Speech, and Language Processing, 27*(4), 730–741.

RStudio Team. (2021). *RStudio: Integrated Development Environment for R* [Available at http://www.rstudio.com/ (Version 1.4.1106)]. RStudio, PBC. Boston, MA.

Ruggles, D., & Shinn-Cunningham, B. (2011). Spatial Selective Auditory Attention in the Presence of Reverberant Energy: Individual Differences in Normal-Hearing Listeners. *Journal of the Association for Research in Otolaryngology, 12*(3), 395–405. https://doi.org/10.1007/s10162-010-0254-z

Sæbø, A. (2001). *Influence of reflections on crosstalk cancelled playback of binaural sound* (Doctoral dissertation). Faculty of Information Technology, Electrical Engineering, Norwegian University of Science, and Technology, Norway.

Sanches Masiero, B. (2012). *Individualized Binaural Technology: Measurement, Equalization and Perceptual Evaluation* (Doctoral dissertation). RWTH Aachen University. Aachen, Germany, Logos Verlag Berlin GmbH.

Sanchez-Vives, M. V., & Slater, M. (2005). From presence to consciousness through virtual reality. *Nature Reviews Neuroscience, 6*(4), 332–339.

Santala, O., Vertanen, H., Pekonen, J., Oksanen, J., & Pulkki, V. (2009). Effect of Listening Room on Audio Quality in Ambisonics Reproduction. *Audio Engineering Society Convention 126*, 1–8.

Savioja, L., Huopaniemi, J., Lokki, T., & Väänänen, R. (1999). Creating Interactive Virtual Acoustic Environments. *Journal of the Audio Engineering Society, 47*(9), 675–705.

Savioja, L., & Svensson, U. P. (2015). Overview of geometrical room acoustic modeling techniques. *The Journal of the Acoustical Society of America, 138*(2), 708–730. https://doi.org/10.1121/1.4926438

Scharrer, R. (2014). *Acoustic field analysis in small microphone arrays* (Doctoral dissertation) [Zugl.: Aachen, Techn. Hochsch., Diss., 2013]. Aachen, Logos-Verl.

Schissler, C., Stirling, P., & Mehra, R. (2017). Efficient construction of the spatial room impulse response. *Virtual Reality (VR), 2017 IEEE*, 122–130. https://doi.org/10.1109/VR.2017.7892239

Schmider, E., Ziegler, M., Danay, E., Beyer, L., & Bühner, M. (2010). Is It Really Robust? Reinvestigating the Robustness of ANOVA Against Violations of the Normal Distribution Assumption. *Methodology, 6*(4), 147–151. https://doi.org/10.1027/1614-2241/a000016

Schmitz, A. (1995). Ein neues digitales Kunstkopfmeßsystem. *Acta Acustica united with Acustica, 81*(4), 416–420.

Schröder, D. (2011). *Physically based real-time auralization of interactive virtual environments* (Doctoral dissertation). Institute of Technical Acoustics, RWTH Aachen University, Germany. Aachen, Germany, Logos Verlag Berlin GmbH.

Schroeder, M. R. (1965). New Method of Measuring Reverberation Time. *The Journal of the Acoustical Society of America, 37*(3), 409–412. https://doi.org/10.1121/1.1909343

Schulgesetz NRW - SchulG. (2021, August 1). Schulgesetz für das Land Nordrhein-Westfalen, §54 Abs. 2 Satz 2 Nr. 1, Landtag Nordrhein-Westfalen [https://recht.nrw.de/lmi/owa/br_text_anzeigen?v_id=10000000000000000524].

Seeber, B. U., Baumann, U., & Fastl, H. (2004). Localization ability with bimodal hearing aids and bilateral cochlear implants. *The Journal of the Acoustical Society of America, 116*(3), 1698–1709. https://doi.org/10.1121/1.1776192

Seeber, B., Kerber, S., & Hafter, E. (2010). A System to Simulate and Reproduce Audio-Visual Environments for Spatial Hearing Research. *Hearing Research, 260*(1-2), 1–10. https://doi.org/10.1016/j.heares.2009.11.004

Sellers, G., Wright Jr, R. S., & Haemel, N. (2013). *OpenGL SuperBible: Comprehensive Tutorial and Reference*. Addison-Wesley.

Sellick, P. M., Patuzzi, R., & Johnstone, B. M. (1982). Measurement of basilar membrane motion in the guinea pig using the mössbauer technique. *The Journal of the Acoustical Society of America, 72*(1), 131–141. https://doi.org/10.1121/1.387996

Seraphim, H. (1961). Über die wahrnehmbarkeit mehrerer rückwürfe von sprachschall (on the perceptibility of multiple reflections on speech sound). *Acta Acustica united with Acustica, 11*(2), 80–91.

Shabtai, N. R., Behler, G., Vorländer, M., & Weinzierl, S. (2017). Generation and analysis of an acoustic radiation pattern database for forty-one musical instruments. *The Journal of the Acoustical Society of America, 141*(2), 1246–1256.

Shafaghat, A., Keyvanfar, A., Lamit, H., Mousavi, S. A., & Majid, M. Z. A. (2014). Open Plan Office Design Features Affecting Staff's Health and Well-being Status. *Jurnal Teknologi (Sciences & Engineering), 70*(7), 83–8.

Shannon, C. E. (1948). A Mathematical Theory of Communication. *The Bell System Technical Journal, 27*(3), 379–423. https://doi.org/10.1002/j.1538-7305.1948.tb01338.x

Shapiro, L. G., & Stockman, G. (2001). *Computer Vision: Theory and Applications.* Prentice Hall.

Shaw, E. A. (1974). The External Ear. In W. D. Keidel & W. D. Neff (Eds.), *Auditory system* (pp. 455–490). Springer. https://doi.org/10.1007/978-3-642-65829-7_14

Shekar, R. C. M. C., & Hansen, J. H. L. (2021). An evaluation framework for research platforms to advance cochlear implant/hearing aid technology: A case study with CCi-MOBILE. *The Journal of the Acoustical Society of America, 149*(1), 229–245. https://doi.org/10.1121/10.0002989

Shield, B. M., & Dockrell, J. E. (2008). The effects of environmental and classroom noise on the academic attainments of primary school children. *The Journal of the Acoustical Society of America, 123*(1), 133–144. https://doi.org/10.1121/1.2812596

Shukla, A., Harper, M., Pedersen, E., Goman, A., Suen, J. J., Price, C., Applebaum, J., Hoyer, M., Lin, F. R., & Reed, N. S. (2020). Hearing Loss, Loneliness, and Social Isolation: A Systematic Review [PMID: 32151193]. *Otolaryngology–Head and Neck Surgery, 162*(5), 622–633. https://doi.org/10.1177/0194599820910377

Simon, L. S., Zacharov, N., & Katz, B. F. (2016). Perceptual attributes for the comparison of head-related transfer functions. *The Journal of the Acoustical Society of America, 140*(5), 3623–3632.

Slater, M., Lotto, B., Arnold, M. M., & Sanchez-Vives, M. V. (2009). How we experience immersive virtual environments: The concept of presence and its measurement. *Anuario de psicología, 40*(2), 193–210.

Soli, S. D., & Wong, L. L. (2008). Assessment of speech intelligibility in noise with the Hearing in Noise Test. *International Journal of Audiology, 47*(6), 356–361.

Søndergaard, P. L., & Majdak, P. (2013). The Auditory Modeling Toolbox. In *The technology of binaural listening* (pp. 33–56).

Spong, M. W., Hutchinson, S., Vidyasagar, M., et al. (2006). *Robot Modeling and Control* (Vol. 3). wiley New York.

Spors, S., Wierstorf, H., Raake, A., Melchior, F., Frank, M., & Zotter, F. (2013). Spatial Sound With Loudspeakers and Its Perception: A Review of the Current State. *Proceedings of the IEEE, 101*(9), 1920–1938. https://doi.org/10.1109/JPROC.2013.2264784

Stenfelt, S. (2012). Transcranial Attenuation of Bone-Conducted Sound When Stimulation Is at the Mastoid and at the Bone Conduction Hearing Aid Position. *Otology & neurotology, 33*(2), 105–114.

Stephens, S. D., & Goodwin, J. C. (1984). Non-Electric Aids to Hearing: A Short History. *Audiology*, *23*(2), 215–240. https://doi.org/10.3109/00206098409072836

Stevenson, J., McCann, D., Watkin, P., Worsfold, S., Kennedy, C., & on behalf of the Hearing Outcomes Study Team. (2010). The relationship between language development and behaviour problems in children with hearing loss. *Journal of Child Psychology and Psychiatry*, *51*(1), 77–83. https://doi.org/https://doi.org/10.1111/j.1469-7610.2009.02124.x

Stevenson, J., McCann, D. C., Law, C. M., Mullee, M., Petrou, S., Worsfold, S., Yuen, H. M., & Kennedy, C. R. (2011). The effect of early confirmation of hearing loss on the behaviour in middle childhood of children with bilateral hearing impairment. *Developmental Medicine & Child Neurology*, *53*(3), 269–274. https://doi.org/https://doi.org/10.1111/j.1469-8749.2010.03839.x

Stienen, J. (2022). *Real-time Auralisation of Outdoor Sound Propagation* (Doctoral dissertation). Institute of Technical Acoustics, RWTH Aachen University. Aachen, Germany, Logos Verlag Berlin GmbH.

Stone, M. A., Moore, B. C., Meisenbacher, K., & Derl eth, R. P. (2008). Tolerable Hearing Aid Delays. V. Estimation of Limits for Open Canal Fittings. *Ear and hearing*, *29*(4), 601–617. https://doi.org/10.1097/AUD.0b013e3181734ef2

Strauss, H. Implementing Doppler Shifts for Virtual Auditory Environments. In: In *Audio engineering society convention 104*. Amsterdam, Netherlands: Audio Engineering Society, 1998, May, 1–18.

Szalma, J. L., & Hancock, P. A. (2011). Noise effects on human performance: A meta-analytic synthesis. *Psychological bulletin*, *137*(4), 682. https://doi.org/10.1037/a0023987

Takeuchi, T., & Nelson, P. A. (2002). Optimal source distribution for binaural synthesis over loudspeakers. *The Journal of the Acoustical Society of America*, *112*(6), 2786–2797. https://doi.org/10.1121/1.1513363

Takeuchi, T., & Nelson, P. A. (2007). Subjective and Objective Evaluation of the Optimal Source Distribution for Virtual Acoustic Imaging. *Journal of the Audio Engineering Society*, *55*(11), 981–997.

Taylor, M. M., & Creelman, C. D. (1967). PEST: Efficient Estimates on Probability Functions. *The Journal of the Acoustical Society of America*, *41*(4A), 782–787. https://doi.org/10.1121/1.1910407

Teather, R. J., Pavlovych, A., Stuerzlinger, W., & MacKenzie, I. S. (2009). Effects of tracking technology, latency, and spatial jitter on object movement. *2009 IEEE Symposium on 3D User Interfaces*, 43–50. https://doi.org/10.1109/3DUI.2009.4811204

VIRTUAL ACOUSTICS. (2021). A real-time auralization framework for scientific research, Institute of Technical Acoustics, RWTH Aachen University [Available at http://www.virtualacoustics.org/, accessed on 2022-01-18].

Theunissen, S. C. P. M., Rieffe, C., Netten, A. P., Briaire, J. J., Soede, W., Kouwenberg, M., & Frijns, J. H. M. (2014). Self-Esteem in Hearing-Impaired Children: The Influence of Communication, Education, and Audiological Characteristics. *PLOS ONE*, *9*(4), 1–8. https://doi.org/10.1371/journal.pone.0094521

Thiele, R. (1953). Richtungsverteilung und Zeitfolge der Schallrückwürfe in Räumen (Directional distribution and temporal sequence of sound reflections in rooms). *Acta Acustica united with Acustica, 3*(4), 291–302.

Thiemann, J., & van de Par, S. (2019). A multiple model high-resolution head-related impulse response database for aided and unaided ears. *EURASIP Journal on Advances in Signal Processing, 2019*(1), 9. https://doi.org/10.1186/s13634-019-0604-x

Thurlow, W. R., & Mergener, J. R. (1970). Effect of Stimulus Duration on Localization of Direction of Noise Stimuli. *Journal of Speech, Language, and Hearing Research, 13*(4), 826–838. https://doi.org/10.1044/jshr.1304.826

Thurlow, W. R., & Runge, P. S. (1967). Effect of Induced Head Movements on Localization of Direction of Sounds. *The Journal of the Acoustical Society of America, 42*(2), 480–488. https://doi.org/10.1121/1.1910604

Tierney, N., Cook, D., McBain, M., & Fay, C. (2021). *naniar: Data Structures, Summaries, and Visualisations for Missing Data* [R package version 0.6.1].

Timmer, B. H. B., Hickson, L., & Launer, S. (2018). Do Hearing Aids Address Real-World Hearing Difficulties for Adults With Mild Hearing Impairment? Results From a Pilot Study Using Ecological Momentary Assessment [PMID: 29956590]. *Trends in Hearing, 22*, 2331216518783608. https://doi.org/10.1177/2331216518783608

Tomblin, J. B., Oleson, J. J., Ambrose, S. E., Walker, E., & Moeller, M. P. (2014). The Influence of Hearing Aids on the Speech and Language Development of Children With Hearing Loss. *JAMA Otolaryngology–Head & Neck Surgery, 140*(5), 403–409. https://doi.org/10.1001/jamaoto.2014.267

Torkildsen, J. v. K., Hitchins, A., Myhrum, M., & Wie, O. B. (2019). Speech-in-Noise Perception in Children With Cochlear Implants, Hearing Aids, Developmental Language Disorder and Typical Development: The Effects of Linguistic and Cognitive Abilities. *Frontiers in Psychology, 10*, 2530. https://doi.org/10.3389/fpsyg.2019.02530

Tsingos, N., Gallo, E., & Drettakis, G. (2004). Perceptual Audio Rendering of Complex Virtual Environments. *ACM Trans. Graph., 23*(3), 249–258. https://doi.org/10.1145/1186562.1015710

Uchanski, R. M., & Sarli, C. (2019). The 'Cupped Hand': Legacy of the First Hearing Aid. *The Hearing Journal, 72*(4), 8–9. https://doi.org/10.1097/01.HJ.0000557743.90033.5c

Van Trees, H. L. (2004). *Optimum Array Processing: Part IV of Detection, Estimation, and Modulation Theory.* John Wiley & Sons.

Venables, W. (1998). Exegeses on Linear Models. *S-Plus User's Conference, Washington DC.*

Vestergaard, M. D., Fyson, N. R. C., & Patterson, R. D. (2011). The mutual roles of temporal glimpsing and vocal characteristics in cocktail-party listening. *The Journal of the Acoustical Society of America, 130*(1), 429–439. https://doi.org/10.1121/1.3596462

Vetterli, M., & Nussbaumer, H. J. (1984). Simple FFT and DCT algorithms with reduced number of operations. *Signal Processing, 6*(4), 267–278. https://doi.org/https://doi.org/10.1016/0165-1684(84)90059-8

Vivek, V., Vidhya, S., & Madhanmohan, P. (2020). Acoustic Scene Classification in Hearing aid using Deep Learning. *2020 International Conference on Communication and Signal Processing (ICCSP)*, 0695–0699. https://doi.org/10.1109/ICCSP48568.2020.9182160

Viveros Munoz, R. A. (2019). *Speech perception in complex acoustic environments: Evaluating moving maskers using virtual acoustics* (Doctoral dissertation). RWTH Aachen University. Aachen, Germany, Logos Verlag Berlin GmbH. https://doi.org/10.18154/RWTH-2019-07497

Von Békésy, G. (1947). A New Audiometer. *Acta Oto-Laryngologica, 35*(5-6), 411–422. https://doi.org/10.3109/00016484709123756

Von Békésy, G., & Wever, E. G. (1960). *Experiments in hearing* (Vol. 195). McGraw-Hill New York.

Vorländer, M. (2020). *Auralization - Fundamentals of Acoustics, Modelling, Simulation, Algorithms and Acoustic Virtual Reality*. Springer.

Wagener, K., Brand, T., & Kollmeier, B. (1999). Entwicklung und Evaluation eines Satztests für die deutsche Sprache. I-III: Design, Optimierung und Evaluation des Oldenburger Satztests (Development and evaluation of a sentence test for the German language. I-III: Design, optimization and evaluation of the Oldenburg sentence test). *Zeitschrift für Audiologie (Audiological Acoustics), 38*, 4–15.

Walden, B. E., Surr, R. K., Cord, M. T., Edwards, B., & Olson, L. (2000). Comparison of benefits provided by different hearing aid technologies. *Journal of the American Academy of Audiology, 11*(10), 540–560. https://doi.org/10.1055/s-0036-1592117

Ward, D. B. (2001). On the performance of acoustic crosstalk cancellation in a reverberant environment. *The Journal of the Acoustical Society of America, 110*(2), 1195–1198. https://doi.org/10.1121/1.1386635

Warren, E., & Grassley, C. (2017). Over-the-Counter Hearing Aids: The Path Forward. *JAMA Internal Medicine, 177*(5), 609–610. https://doi.org/10.1001/jamainternmed.2017.0464

Wefers, F. (2015). *Partitioned convolution algorithms for real-time auralization* (Doctoral dissertation). Institute of Technical Acoustics, RWTH Aachen University. Aachen, Logos Verlag Berlin GmbH.

Wefers, F., & Vorländer, M. (2018). Flexible data structures for dynamic virtual auditory scenes (D. Ballin & R. D. Macredie, Eds.). *Virtual Reality*. https://doi.org/10.1007/s10055-018-0332-9

Wendt, K. (1963). *Das Richtungshören bei der Überlagerung zweier Schallfelder bei Intensitäts-und Laufzeitstereophonie (Directional hearing given superposition of two sound fields based on intensity and time-of-arrival stereophony)* (Doctoral dissertation). Rheinisch-Westfälische Technische Hochschule Aachen.

Wenzel, E. M. (1995). The relative contribution of interaural time and magnitude cues to dynamic sound localization. *Applications of Signal Processing to Audio and Acoustics, 1995., IEEE ASSP Workshop on*, 80–83. https://doi.org/10.1109/ASPAA.1995.482963

Wenzel, E. M., Arruda, M., Kistler, D. J., & Wightman, F. L. (1993). Localization using nonindividualized head-related transfer functions. *The Journal of the Acoustical Society of America*, *94*(1), 111–123. https://doi.org/10.1121/1.407089

Westermann, A., & Buchholz, J. M. (2015). The influence of informational masking in reverberant, multi-talker environments. *The Journal of the Acoustical Society of America*, *138*(2), 584–593. https://doi.org/10.1121/1.4923449

Whittaker, E. (1915). On the functions which are represented by the expansion of interpolating theory. *Proc. Roy. Soc. Edinburgh*, *35*, 181–194. https://doi.org/10.1017/S0370164600017806

Wichmann, F. A., & Hill, N. J. (2001). The psychometric function: I. Fitting, sampling, and goodness of fit. *Perception & psychophysics*, *63*(8), 1293–1313. https://doi.org/10.3758/BF03194544

Wightman, F. L., & Kistler, D. J. (1997). Monaural sound localization revisited. *The Journal of the Acoustical Society of America*, *101*(2), 1050–1063. https://doi.org/10.1121/1.418029

Wightman, F. L., & Kistler, D. J. (2005a). Measurement and Validation of Human HRTFs for Use in Hearing Research. *Acta Acustica united with Acustica*, *91*(3), 429–439.

Wightman, F. L., & Kistler, D. J. (2005b). Informational masking of speech in children: Effects of ipsilateral and contralateral distracters. *The Journal of the Acoustical Society of America*, *118*(5), 3164–3176. https://doi.org/10.1121/1.2082567

Williams, E. G. (1999). *Fourier Acoustics: Sound Radiation and Nearfield Acoustical Holography*. Academic Press.

Williams, E. (1949). Experimental designs balanced for the estimation of residual effects of treatments. *Australian Journal of Chemistry*, *2*(2), 149–168. https://doi.org/10.1071/PH490149

Wingfield, A., Tun, P. A., & McCoy, S. L. (2005). Hearing Loss in Older Adulthood: What It Is and How It Interacts With Cognitive Performance. *Current Directions in Psychological Science*, *14*(3), 144–148. https://doi.org/10.1111/j.0963-7214.2005.00356.x

World Health Organization. (2021, April). Deafness and hearing loss - Fact sheet [Accessed on 2021-08-27].

Wright, D., Hebrank, J. H., & Wilson, B. (1974). Pinna reflections as cues for localization. *The Journal of the Acoustical Society of America*, *56*(3), 957–962. https://doi.org/10.1121/1.1903355

Xie, B. (2013). *Head-related transfer function and virtual auditory display*. J. Ross Publishing.

Yairi, S., Iwaya, Y., & Suzuki, Y. (2006). Investigation of System Latency Detection Threshold of Virtual Auditory Display. *Proceedings of the 12th International Conference on Auditory Display*, 217–222.

Yellamsetty, A., Ozmeral, E. J., Budinsky, R. A., & Eddins, D. A. (2021). A Comparison of Environment Classification Among Premium Hearing Instruments [PMID:

33749410]. *Trends in Hearing*, *25*, 2331216520980968. https://doi.org/10.1177/2331216520980968

Yoshinaga-Itano, C., Sedey, A. L., Coulter, D. K., & Mehl, A. L. (1998). Language of Early- and Later-identified Children With Hearing Loss. *Pediatrics*, *102*(5), 1161–1171. https://doi.org/10.1542/peds.102.5.1161

Yuen, K. C. P., & Yuan, M. (2014). Development of Spatial Release From Masking in Mandarin-Speaking Children With Normal Hearing. *Journal of Speech, Language, and Hearing Research*, *57*(5), 2005–2023. https://doi.org/10.1044/2014_JSLHR-H-13-0060

Zahorik, P. (2002). Assessing auditory distance perception using virtual acoustics. *The Journal of the Acoustical Society of America*, *111*(4), 1832–1846. https://doi.org/10.1121/1.1458027

Zahorik, P., Brungart, D. S., & Bronkhorst, A. W. (2005). Auditory Distance Perception in Humans: A Summary of Past and Present Research. *ACTA Acustica united with Acustica*, *91*(3), 409–420.

Ziegelwanger, H., Majdak, P., & Kreuzer, W. (2015). Numerical calculation of listener-specific head-related transfer functions and sound localization: Microphone model and mesh discretization. *The Journal of the Acoustical Society of America*, *138*(1), 208–222. https://doi.org/10.1121/1.4922518

Zotter, F. (2009a). *Analysis and synthesis of sound-radiation with spherical arrays* (Doctoral dissertation). Institute of Electronic Music and Acoustics, University of Music and Performing Arts Graz, Austria.

Zotter, F. (2009b). Sampling Strategies for Acoustic Holography/Holophony on the Sphere. *NAG/DAGA 2009 International Conference on Acoustics, Rotterdam*, 1107–1110.

Zotter, F., & Frank, M. (2012). All-Round Ambisonic Panning and Decoding. *Journal of the Audio Engineering Society*, *60*(10), 807–820.

Zotter, F., & Frank, M. (2019). *Ambisonics. A Practical 3D Audio Theory for Recording, Studio Production, Sound Reinforcement, and Virtual Reality*. Springer International Publishing. https://doi.org/10.1007/978-3-030-17207-7

Zotter, F., Frank, M., & Pomberger, H. (2013). Comparison of energy-preserving and all-round Ambisonic decoders. *Fortschritte der Akustik, AIA-DAGA,(Meran)*.

Zwicker, E., & Fastl, H. (2013). *Psychoacoustics: Facts and models* (Vol. 22). Springer Science & Business Media.

Acknowledgements

I would like to take this opportunity to express my sincere gratefulness to the people without whose support this work would not have been possible.

First of all, I would like to thank Univ.-Prof. Dr.-Ing. Janina Fels for providing me with the opportunity to pursue my PhD at this renowned institute of acoustics, the uncomplicated supervision, and the granted freedom to conduct my research. Professor Astrid van Wieringen, PhD, kindly took over the co-examination of my thesis. She also initiated and efficiently coordinated the project this journey started with: iCARE was just a great experience!

My sincere respect goes to Univ.-Prof. Dr. rer. nat. Michael Vorländer who managed to maintain a relaxed and connected research environment. Michael, I am grateful for your helpful comments and advice, and really enjoyed the few business trips together ("fusca")! Fortunately, I had the pleasure of experiencing the institute's everyday life in the presence of Dr.-Ing. Gottfried Behler who always had an open ear for all kinds of acoustic problems. Thank you, Gottfried, for sharing your extensive knowledge and helping to solve various challenges!

Essential support also came from my colleagues: Special thanks go to Dr.-Ing. Johannes Klein, my first office colleague, who introduced me to the institute and the ITA-Toolbox, and disseminated his expertise in signal processing and acoustic measurements. Jonas Stienen, my second office colleague, became one of my best friends. Thank you for the maintenance and continuous improvement of VA, your input to own implementation problems, and much more. I will never forget the great times we had at conferences and in our free time, also with your family. The close and cordial cooperations with Michael Kohnen and particularly Dr.-Ing. Lukas Aspöck were very instructive and demonstrated that hard work and fun can be easily combined. Kudos to Marco Berzborn for all the helpful discussions and comments, as well as the technical support. Mark Müller-Giebeler kindly provided input to handle the conversion of the caravan and was always

helpfully available. Thanks, Mark, for finding the time to play tennis despite your
busy schedule. I also appreciate the collaborations with and support of Zhao Ellen
Peng, PhD, Dr.-Ing. Ramona Bomhardt, Shaima'a Doma, Dr. Manuj Yadav, Hark
Braren, Dr. rer. nat. Frank Wefers, Sönke Pelzer, Dr.-Ing. Markus Müller-Trapet,
Dr.-Ing. Martin Guski, Dr.-Ing. Martin Pollow, Rob Opdam, Dr.-Ing. Fanyu
Meng, and Dr.-Ing. Noam Shabtai.

Since some of the presented experiments strongly depended on hardware,
the mechanical and electronic workshops played crucial roles. Hats off to Uwe
Schlömer, Mark Eiker, and Thomas Schaefer, who always implemented my detailed requests in a very timely manner at a level that is second to none. Rolf
Kaldenbach was in no way behind this level of professionalism, particularly when
installing the high-channel loudspeaker array in the anechoic chamber. Thank
you, Rolf, for your invaluable help! I would like to extend my appreciation to
Karin Charlier for making the organisational processes as simple as possible,
handling financial matters, and communicating things in a very congenial way,
and Norbert Konkol for troubleshooting network problems.

It was also a pleasure to work with some very skilled students: Anne Stockmann, Jonas Tumbrägel, Lukas Vollmer, Karin Loh, Wenzel Neuhöfer, and Lena
Wirtz provided important contributions either to this thesis or other projects,
and continuously challenged me to improve my knowledge. The support during the organisation of experiments and data collection by Laura Bell, PhD,
Dr. rer. medic. Lucia Martin, Alokeparna Ray, Cosima Ermert, and Arndt Brandl
is highly appreciated, as is the help of Julian Burger, Robert Henzel and Luiz
Otavio Kohler. External collaborations enabled the completion of project mile-
stones in the first place, others opened up new possibilities: Many thanks go to
Univ.-Prof. Dr. med. Christiane Neuschaefer-Rube from Uniklinikum Aachen,
Frank Keller-Drees from Hörzentrum Euregio Aachen, Judith Verberne, Matthew
McKee and Todd Fortune, PhD, from GN ReSound, Per Blysa from Sonion,
Carsten Svensson and Michael Kissel from Ecophon, and Prof. Dr.-Ing. Stephan
Paul from UFSC Florianópolis, Brazil. I also appreciate the collaborations with
my colleagues within the iCARE project and the staff from the Department of
Environmental Psychology at the Högskolan i Gävle, Sweden.

To my friends in Aachen, I would like to express my heartfelt thanks for
distracting me (very efficiently) from my daily work routine. I will especially
miss the regular activities with Jens Mecking, Dr. Conrad van der Laden, Tristan
Heider, and Marius Heidweiler, and their families.

Last but not least, my utmost gratitude goes out to my family. In particular,
I am deeply indebted to my parents Brigitte and Johannes for their unconditional
support throughout my life. Without you, I would not have come that far. This
thesis is dedicated to you.

Bisher erschienene Bände der Reihe
Aachener Beiträge zur Akustik

ISSN 1866-3052
ISSN 2512-6008 (seit Band 28)

Alle erschienenen Bücher können unter der angegebenen ISBN-Nummer direkt online (http://www.logos-verlag.de) oder per Fax (030 - 42 85 10 92) beim Logos Verlag Berlin bestellt werden.